Trading Paces

MICHAEL KELLY worked for ten years in the IT industry in Dublin selling computer 'stuff'. Then he and his wife abandoned their Celtic Cub, commuter-hell lifestyle for rural idyll in the County Waterford countryside, living in a leaky cottage on an acre, growing vegetables (badly) and rearing a few animals for the table. He now works as a writer and freelance contributor to *The Irish Times* and *The Gloss* magazine. He has also appeared on 'The Green Light', 'The Dave Fanning Show' and 'Mooney' on RTÉ Radio 1. In 2005 he recorded an album, *Amongst Women* — which, lamentably, nobody bought.

Trading Paces

From Rat Race to Hen Run

Michael Kelly

THE O'BRIEN PRESS
DUBLIN

First published 2008 by The O'Brien Press Ltd,
12 Terenure Road East, Rathgar, Dublin 6, Ireland.
Tel: +353 1 4923333; Fax: +353 1 4922777
E-mail: books@obrien.ie
Website: www.obrien.ie

ISBN: 978-1-84717-070-5

British Library Cataloguing-in-Publication Data
Kelly, Michael
Trading paces : from rat race to hen run
1. Kelly, Michael 2. Self-reliant living - Ireland 3. Farms, Small - Ireland 4. Country
life - Ireland
I. Title
630.9'2

1 2 3 4 5 6 7 8 9 10

08 09 10 11 12 13 14 15

Layout and design: The O'Brien Press Ltd.
Printed and bound in the UK by J.H. Haynes & Co. Ltd, Sparkford.

Dedication

To the long-suffering and ever-patient Mrs Kelly –
none of this would have been possible without you.

Acknowledgements

Huge thanks to all of the following for their help and support over the past few years. My mum – the reason we gravitated to Dunmore in the first place. My sisters – Kim, Niamh, Ciara and Dara – and their respective families. My extended family in Ballygalvert – in particular the father-in-law for sage advice, the mother-in-law (hen supplier-in-chief and an inspiration when it comes to growing veg) and Stephen, for very practical assistance on gardening projects too numerous to mention.

I was given a great deal of inspiration, advice and encouragement by many friends and colleagues (past and present). If I try to list them I will invariably leave someone out, so thank you all. Our neighbours and friends here in Dunmore East, particularly Nathaniel and Ruth. To Denis Shannon and Betty Byrne for all the support and advice with the Tamworth Two. To a very talented photographer, Seamus O'Neill – just three words of advice: 'It's your calling.' Patsey Murphy in *The Irish Times* and Sarah McDonnell in *The Gloss* showed a faith in my abilities without which I might never have left the IT industry – for that I owe them a debt of gratitude. Also thanks to Kate Holmquist, Clare McCarthy, Carmel Daly, Liam McAuley, Conor Goodman, Barry O'Keefe and Jane Powers in *The Irish Times*. Finally, to Michael O'Brien, and all at The O'Brien Press for taking a punt on *Trading Paces*, and Ide ní Laoghaire, my editor, for guiding things along so expertly.

CONTENTS

PREFACE

I have this thing I call the Wedding Test which goes like this: you are sitting at the reception of a wedding and making small talk with the person beside you while tucking in to your slab of beef (or salmon). You have discussed the bride and groom, the bridesmaid's dresses, the mother-of-the-bride's hat, how nice the service was, the table centrepiece, how ridiculously OTT weddings have become ('They spent €40,000?!! My God, are you kidding me?!'), how nervous the best man is about his speech, the institution of marriage, kids, traffic, the price of property. Having established a wine-fuelled comfortable rapport, your dinner companion feels comfortable enough with you to pop the question:

'So, eh, what do you do?'

Up to very recently I've always dreaded that moment. It's a loaded question, a question that judges you. Measures you. A question that *expects*. 'What do you do?' Four little words with a sting in the tail. Four little words, the answer to which will define you in the eyes of that person. If, for example, you say that you're a managing director or a doctor, this will immediately establish you as a Successful Person in your dinner-companion's eyes. Yes, we are all just that shallow.

Now, how do *you* feel about the answer that comes out of your mouth? Are you happy with your response? Are your proud of what you do for a living? Or at least comfortable with it? Or do you mutter something inaudible through a mouthful of roast potato and hope someone will start clinking glasses to herald the start of the speeches?

For ten years I did the latter. 'I'm a salesman,' I would mumble and hope that they didn't hear me or that the person on the far side of them would ask a question and they would be distracted. Somewhere else at the table Mrs Kelly (my wife is an insanely private person and has therefore expressly forbidden me to use her first name, Eilish, anywhere in this book) would be saying, 'I'm an accountant' to *her* dinner companion and watching their eyes glaze over.

More often than not I would be too embarrassed to say, 'I'm a salesman,' and would say instead, 'I work for an IT company.' Then they would say, 'Oh very good. Are you a programmer?' And I would say, 'Eh, no, I'm in sales.' I tried to avoid the word 'salesman' as much as I could. I can't really explain why I was so embarrassed by that word – and I *was* more embarrassed by the word than by the profession itself. The job itself I could just about handle. But the word 'salesman' – I hated that word. (Incidentally, I'm not alone – you will never find a salesperson with the word 'salesperson' or 'salesman' on their business card. It will say 'account manager', 'sales executive', 'new business development manager', 'sales account executive', 'sales development manager'. Pretty much *anything* but 'salesperson').

To many people the word 'salesman' conjures up an image

of someone slightly shifty, perhaps even a bit false or smarmy. It always seems to suggest that something unpleasant is about to happen. If you ask someone what a salesperson will be like socially, they will say that they might be fun and there certainly wouldn't be too many lulls in the conversation (sales people do like to talk, after all) – but won't they always be looking for an angle? Can we trust them, really? We've become incredibly tired and cynical about sales, advertising, marketing and PR – and rightly so, in my view. We always tend to think that people in those professions are trying to con us into buying products we don't need and, worse, they have Jedi mind-tricks up their sleeves that make us powerless to resist their charms.

Back at the dinner table, if the conversation developed beyond these initial forays, and it rarely did, I would find myself getting all defensive about my job, quite uninvited. I would say things like, 'Well, sure, it's only a job' or 'It pays the bills.' But, of course, a job is never only a job and we're deluding ourselves when we say this. It's far more than that. In the average working life of forty-five years, we will spend 10,800 days at work. Think about that: 10,800 days. Sounds a lot, doesn't it? Our job is something we will be doing until we're sixty-five – so we'd better bloody well like it.

And I was starting to realise – all too slowly – that I bloody well didn't.

Chapter 1

RETIREMENT AS THE ULTIMATE GOAL IN LIFE

I've never understood why as a society we have all bought into the notion of retirement being the ultimate goal of a working life. It's held up as this utopian destination that we'll reach one day in the distant future when we'll be able to kick back and enjoy life. In the meantime we are willing to put up with all manner of misery in the workplace. 'My job's shit, but never mind – when I get to sixty-five, *then* I'll live.' My father died when he was fifty-seven and I think that has always made me circumspect about the chances of even reaching retirement age. Imagine working away, thinking, 'Ah, at least in eight years' time I'll be able to retire, then I'll live, goddammit!' and then, kablamo, your time is up. You'd be pretty miffed, wouldn't you?

I remember sitting in a meeting with a pensions adviser when I was about twenty-five and he told me I had to start putting IR£400 a month into my pension fund – I had to stop myself from laughing in his face. The notion that I would deliberately deprive myself of such a large chunk of my earnings and let some bank hold onto it for the next forty years just seemed absurd to me, particularly given that I might not be around to get it back.

In your twenties you don't tend to waste much time thinking longterm. In fact, if you think at all, it's probably about sex or beer, and how to get your hands on lots and lots of sex and beer. You tend to 'work to live' as opposed to 'live to work' – a tired old cliché, but there's a lot of truth in it. You work to get money to buy beer which will hopefully lead you eventually to more sex. It's a simple formula. Your twenties was a decade for pubs, clubs, parties, sexual exploration and falling down drunk. Your thirties is a decade for getting real.

When I turned thirty I started to think long and hard about where I was going with my working life and whether I wanted to spend any more time as a salesperson. I remember reading an obituary in a newspaper one day and wondering to myself: when *I* die, will there be an obituary in the paper? I know that sounds painfully vain – and don't get me wrong, I know that (a) there are loftier measures of our worth as human beings then whether we have done something newsworthy and (b) there are also other things in life that are far more important than our careers – the type of person we were, the way we treated others and so on. But you get my point. If you can wangle an obituary it probably means you have done something of note in your life. Would anyone want to write an obituary about a salesman? *'Michael Kelly died yesterday aged 57 while in the middle of a sales pitch on storage area networks to ABC Corp. He was on slide 18 of a 138-slide PowerPoint presentation when he keeled over dead from boredom.'*

In your thirties you are more than likely settled down, hopefully with the love of your life, so when it comes to sex there's a status quo of sorts in operation – you've a fair idea when and how regularly it will arrive so you stop getting worked up about it in the interim. As for beer, well, for some inexplicable reason these days, two or three pints makes you slur your words and fall asleep into your pint and your hangovers now seem to last four days. You start using a phrase that you thought would never pass your lips: 'I can't drink any more.' Most importantly, in your thirties you can do something which you could not do in your twenties: you can imagine yourself being forty. Or fifty. Or, God forbid, old and wrinkly. You can imagine yourself being un-cool, driving too slowly, befuddled, overly fond of routine and finding music too loud. You realise, with considerable anguish, that people in their teens and twenties now consider you to be a stuffy old fool, even though you think you're still pretty hip.

I remember in my twenties being convinced I was going to be a millionaire. It wasn't a childish pipedream – it was an unshakeable belief, so much so that I voiced it to other people, most notably the future Mrs Kelly back when she was my girlfriend. In fact, I think I actually wrote her a cheque for a million pounds when I got my first chequebook, which, if I am not mistaken, she still has (that's a little freaky, come to think of it, but she was an accountant after all). I wasn't sure *how* I was going to become a millionaire – I had no actual plan to achieve this grand financial milestone. As I frittered away my twenties selling IT systems I could console myself that I had lots of time to switch to that new career or stumble on that great invention or idea which would bring me my millions and allow future Mrs Kelly to cash that cheque. But in my thirties I started gradually to realise: you know what, that's not going to happen, and that gurgling noise that I could hear in my sub-conscious was the sound of the last vestiges of my youthful dreams going down the plughole.

For the first time I realised that this job I was in might actually be what I would spend the rest of my life doing. That thought alone pitched me into sheer terror. You build up capital when you spend time in a job. You build up expertise, knowledge, credit, recognition, maybe even renown. All of which makes it difficult to jump ship to another career. So you don't. You pass five years in a job, then seven, then ten. Pretty soon it's twenty, then thirty years. Next thing you know, a young CEO is thanking you for your long years of service and handing you a gold watch and a card signed by colleagues, most of whom don't know your name. And then? Apparently, that's when you're supposed to start having fun.

I couldn't imagine being a salesman at forty years of age, never mind sixty-five, and once you arrive at that realisation, moving on becomes an absolute necessity. Thankfully, it was made less difficult by the fact that I was gradually becoming disillusioned by the corporate world and the grubby sciences that surround it: sales, marketing, PR, advertising, guff, lies, pretend, bend-a-rule, break-a-rule, *tell the customer what they want to hear*. In my last five years working in sales I built up what I consider to be a healthy suspicion about the corporate motive. When it comes to the crunch, capitalism is cold and brutal, and its overriding loyalty is delivering profit to shareholders. Companies don't really give a rats about the people they employ. We are expendable. We are a means to an end. Most of us know this but we forget at times in our lives and think the company we work for is somehow different. Or we think that *we* are somehow different – better and therefore immune.

My reasons for staying put for nearly ten years in a job I disliked were varied and complex. Firstly there was that general malaise and fear of the unknown that cripples us all to some degree. The fact that I was earning really good money was also a factor – good wages are very seductive and it's tough to deliberately turn off the tap. Bizarrely, earning good money tends to tie you down – 'I'm earning good money,' you tell yourself, 'so now's a good time to borrow!' I seemed to have accumulated a dizzying array of fiscal responsibilities – mortgage, loans, credit-card bills. You can't just give up your job and live on air, can you? And anyway, even if I did move on, what would I move on to? Sales is quite specialised. People think that a salesman is skilled at selling and can sell anything. But that's not quite true. What I was good at was selling IT systems; I could no more move to selling, say, chemicals, than I could move to being a racing driver. I could move to a different IT company, but that would just be the same shit, different venue. So on Fridays I would open the jobs supplement believing firmly that some new career, some ideal job would magically pop out from the page, delivering me from my purgatory. But it never happened, probably because I didn't have any clue what the ideal job was.

There's a saying that if you can work at something you love, you will never work a day in your life. It sounds wonderful, but, I wondered, are there really people out there for whom that holds true? When I thought about it, which I did frequently, there were two possible jobs that I considered would be my dream jobs: something to do with music or something to do with writing. Both of those things were hobbies of mine. But just how do you go about turning a hobby into a job, what you might call a jobby? So I stayed put. And the years rolled past.

When I look back on it now, I feel pretty mad that I wasted so many years whining and moaning and doing damn all to change my situation. I'm convinced that there is one reason we settle down, comfortable but unhappy, in the little rut we scratch out for ourselves: consumerism. Consumerism is a vast, highly complex conspiracy. It keeps us wedged in jobs we don't like, working every day God sends so that we can keep paying for the cobble-lock driveway, the iPods, MP3 players, spanking new Mini Cooper convertible, hot-tubs, a collection of DVDs we will never watch and CDs we will rarely listen to, 55-inch, high-definition plasma TVs, gym sub-scriptions, pilates programmes, yoga holidays. As a wise man once said, we are more eager to amass than to realise.

The real genius of this conspiracy is that we remain completely oblivious to the fact that if we simplified our lives slightly and got rid of (or didn't buy) the stuff we don't really need we mightn't have to work as hard or for as long. Or even more intriguing, we could choose to work at something we love even if it didn't pay so well. The conspiracy has been so successful it has warped the nature of work itself beyond all recognition. Work is no longer a means to an end; it is no longer a way to provide food and shelter for ourselves and our loved ones. It has become, instead, something we MUST spend an entire lifetime doing and all the while the original objective, ie providing food and shelter, becomes a sort of by-product. If we don't buy into this model, we are considered slackers, wasters, losers. We are not game-players. Work has become so all-encompassing it has relegated *living our lives* to second place.

This book doesn't pretend to know how to unravel that complex conspiracy completely or even slightly. All it shows is how two people shouted ENOUGH and then started to simplify, to downshift. To opt out. If we can claim any credit at all, which we don't, it is that we finally came to a really small but very important (and entirely obvious) realisation: less money going out means you need less money coming in. And that opens up a whole world of opportunities. All of which is really easy to say, fairly easy to type, but very hard to do. It means flipping everything you know on its head.

Economists tell us that cheaper credit has liberated Irish people. When interest rates were 18 percent or whatever they were in the bad old days, people could only afford a house worth IR£20,000 and the repayments still swallowed up half their salaries. These days interest rates are as close to 0% as they can be (although they are creeping up) so we can *afford* to spend €800,000 on a two-bedroomed townhouse in some previously rural town about two hours' drive from work, can't we? One hundred percent mortgages are almost the norm so there's not even a necessity to save a deposit these days. Whenever any pundit pops up on TV or radio to offer the possibility that this might not be the greatest idea in the world, they get shot down as if just by mentioning it they are not playing ball. *Shame on you, Sir! Cheap credit has made us free!*

Arse! Cheap credit has made us slaves! Hundreds of thousands of couples locked into thirty- and forty-year mortgages with repayments that are so high there will never be any other option but for both of them to work all the hours God sends. Forty years! What age will they be when they finally get their mortgage paid off? Sixty-five. Ah! Retirement age – the Garden of Eden, El Dorado, Nirvana, the old utopian chestnut, the pot o' gold at the end of the rainbow. An entire working life dedicated to paying off a bank that makes more money in a minute than you would in a hundred lifetimes. And because both parents have to work to pay off the mortgage, the little darling they brought into the world gets plonked into a crèche from dawn to dusk. You're not happy about it, but what other option do you have? The exorbitant costs of childcare lock down the rut even further until there is literally no way out of it. That's not freedom. It's a modern form of slavery.

Part of the problem is the desire to *amass,* to have stuff. We discovered that it's only when you accept that as fact, that you can start to consider downshifting as an option. Our aspirations too keep us trapped just as surely as a monthly mortgage repayment and in some ways they are harder to let go of. Aspirations can be just as tight-fitting a straitjacket as *things.* Downshifting isn't easy in modern society because it is all about accepting that less is more. It's about making sacrifices. It's about saying: I do not want a house as big as our neighbours. I do not want a new car. I don't want three holidays a year. In modern Ireland, which is fundamentally about wanting pretty much *everything* and where our value as a person is measured by how much stuff we own, downshifting is also about ploughing a lonely furrow.

Now, I began to say to myself, what if you worked at something you love and you don't *want* to retire when you're sixty-five? Imagine the freedom that would grant you? Then you wouldn't need to be giving €500 a month to the pensions people. And on your spreadsheet where you mark down all your incomings and outgoings, you can take €500 out of the outgoings column. And that's when I began to see the first chink of light.

Well, if I need €500 less each month, I thought, then I can afford to earn €500 less. *Right now.* And maybe instead of working twelve hours a day I can work, let's say, eight or seven or six. Or maybe four days a week instead of five. The extra free time would be sort of like moving retirement forward by thirty-five years or so. Which makes sense because at least I know that now I am blessed with the good health to enjoy it. Now I was starting to see the potential.

I began to wonder was there anything else I could take out of the outgoings column. There was a car loan for €400 a month. It's a hell of a nice car. But how much free time would €400 a month buy me, if I was to sell it off and buy something cheaper? I was making progress.

But I needed to take stock. What was I good at? How could I begin to make these changes? Could I make a living from a jobby?

Chapter 2

HELP, HELP!
THE SKY IS FALLING!

In my mid-twenties I tried writing a book in my spare time and sometimes (whisper it) at my desk at work. It was called *The First-time Buyer's RANT.* We were buying our first house at the time and basically the book was a long series of complaints about all the stuff that first-time buyers have to put up with: never-ending snag-lists, dodgy builders, bone-idle solicitors, absentee electricians, bathroom-tile salespeople and so on. The title, a clever if rather obvious pun on the first-time buyer's grant, was the best thing about the book and it all went downhill from there. Undeterred, I printed off a few chapters and sent them to some publishers, most of whom studiously ignored it. One well-known Irish publisher gave it a proper look, sending me into a brief spin of nervous excitement, but eventually they sent me a rejection slip. That's the only thing JK Rowling and I will ever have in common.

The year I turned thirty I decided that my deliverance was to be found in the music industry, more specifically by releasing an album. Clearly, this was the manifestation of an early-onset midlife crisis. Some men buy a Porsche or have a tempestuous affair with Jenny from Accounts – I convinced myself I was going to be a rock star. It sounds gloriously self-delusional now, but in my twenties I did earn some money singing and playing guitar in pubs. Though quite what made me think I had what it took to make it big in the music business is another story. Anyway, there was a flurry of creative activity and a few months later I had written about twenty songs – they were mostly crap, but hey, at least I was prolific. That summer I spent my two-week holiday holed up in a record producer's gaff recording the songs and subsisting on a diet of tea, crisps and McVitie's Boasters. I'd like to say that I embraced the rock 'n' roll lifestyle, but in fact I found it sort of irritating. Would it have killed him to lay on a few decent sandwiches or a bowl of fruit, for example? Our views on what constitutes a working day were also hopelessly out of sync – he wouldn't get out of bed until the afternoon and would work until the wee hours of the morning, which didn't suit me at all. Sure, I wanted to release a hit album – but I wanted to do it during business hours (I was still a corporate man then). And then there was all that tobacco he was smoking – had he ever heard of the smoking ban in the workplace, I wondered, gasping in desperation? The album was produced on a shoestring – the producer wanted a full sound on all the songs but we couldn't afford to hire musicians, so we played all the instruments between the two of us and production techowizardry did the rest. If Mrs Kelly was bothered by the fact that her husband was off blowing the annual holiday budget on a misguided musical opus, she never let on.

On the album cover there's a cringe-making photo of me sitting amongst three mannequins in a shop window (a long story) looking decidedly uncomfortable – perhaps I was aware even then that thirty-year-old salesmen have no place on the front cover of an album and that I needed to cop myself on a small bit. I got a hundred copies pressed and even set up a website where you could buy the album and listen to sample tracks. Strangely, once I had all that done, I sort of lost interest. I sent off some copies to record companies and radio stations, half-expecting a limo to pull up at the front door at any moment to whisk me off to my new life as a superstar and half-knowing that this wouldn't happen. The next steps should have involved the hard graft that all successful artists have to go through to get their big break – knocking on doors, gigging every two-bit venue in the country etc. But my heart just wasn't in it. One DJ on Today FM, AM Kelly (no relation, incidentally), took one of the songs, 'Mr Brown's War', to heart and gave it some airplay (well, she played it twice, which was hardly high-rotation). I also played a gig in Whelan's in Dublin, but everyone in the audience was either a friend or a family member – there may have been one other guy there too but I think he was lost. Those brief forays into stardom were to be the high point of a short-lived musical career. Feeling pretty dejected that the world at large seemed spectacularly disinterested in my debut book *and* my debut album, I went back to selling.

Thankfully, while I was faffing about writing books that no one wanted to read and releasing albums that no one wanted to listen to, Mrs Kelly was busy making real plans to improve our lot in life. She called me one day from work to say that she had been on the internet and had found an old cottage in the countryside that she wanted us to go see. I should explain that after living in sin together for years in apartments and flats in Dublin, early married life and economic realities saw us buying a house in Gorey, County Wexford, which was about as close to the capital (eighty kilometres away) as we could afford. These days, Gorey is practically a suburb of Dublin, but back then we and our fellow estaters blazed a bit of a trail in terms of having such a long commute to work. People thought we were nuts when we told them that we were traversing six counties a day just to get to work. That's thirty counties a week – almost a united Ireland.

At that stage we were both working for companies in Dun Laoghaire so we used to commute together, which lightened the load somewhat. We had plenty of what relationship counsellors call 'quality time' together. The journey was a tough old slog, but we just got on with it and if people said, 'God, that must be tough', we would actually defend the lunacy of a three-hour daily commute. We would say things like, 'Ah it's not too bad actually, it's not a bad road' or 'We're lucky, we share the driving', or my favourite: 'I actually like driving.' It took us two years to realise that it's not really *living* in the proper sense of the word to spend fifteen hours a week driving to your job.

Amazingly, a whole generation of us have come to accept this as the norm – I heard a guy on the radio recently who commutes from Kilkenny every day to Dublin. It takes him over two hours EACH WAY. That's four hours a day, twenty hours a week, about forty days a year spent in the car getting to work. Imagine how things were a hundred years ago for people in Kilkenny. They might have got to Dublin once a year, or once in a lifetime. It would have been an excursion, a trip, an event. It would have been planned for and taken seriously. Sandwiches would have been made, stop-offs planned, rugs brought along to drape over frigid knees. Now, people do it every day – just to get to work. It's crazy.

Our first house was the quintessential Celtic Tiger dwelling – a three-bed semi-D in a sprawling estate on the outskirts of a town on the southernmost extremity of the Dublin commuter belt. It was a grand little house – there are thousands the very same in every town in Ireland. All the little culs-de-sac in the estate had ridiculous horsey names designed to try and evoke the countryside: The Canter, The Gallops, The Neigh, The Trots.

When you don't know anyone in an estate it can be a terrifyingly soulless place to live. Full as it was of commuters, most of the houses stood empty for twelve hours a day from Monday to Friday and when we were there we were either lying half-dead on the sofa, or sleeping, eating and catching up on laundry. We tried to blend in with the local community as best we could, but we had no connection with Gorey whatsoever. We knew no one and given that we were mainly too tired to socialise, there was little prospect of us ever getting to know anyone.

We got up at 5:30 each morning to be on the road shortly after 6:15. Those forty-five minutes were a mad scramble to get showered and shaved (Mrs Kelly didn't shave obviously), have breakfast, drag our poor dog for a bleary-eyed walk around the estate, and then get out the door. We usually tried to get away from work by about 4:30pm to compensate for the fact that we had an hour and a half of work done by the time the rest of our co-workers came in at nine. But you still always got 'looks' from people when you were walking out of the office and almost every day, EVERY BLESSED DAY, someone would feel compelled to point at their watch and ask 'Half day?' as you walked to the door. It's hard to explain to someone who has never had a long commute how unspeakably grim it is to face into a one-and-a-half-hour drive having got up at 5:30am and put in a full day's work. It wouldn't be so bad if you only had to do it every now and then – but to do it each and every day, week-in, week-out, was soul-destroying. We went to bed at about nine o'clock each evening, completely shagged out. I remember once in the height of summer being really tempted to get up and shout something abusive out the window at kids playing on the green in front of our house as we lay in bed trying to get to sleep. Then I looked at the bedside clock and realised it was only 8.30 and it was still broad daylight.

Eventually, both of us fought for concessions at work in terms of working from home and, in fairness to both companies, they bought into this. When your boss allows you to work from home one or two days a week, he does, of course, reserve the right to make you feel like he's doing you an enormous fecking favour and co-workers will have some spiteful fun with it too: 'Ah! working from home, huh? God, that must be great. I bet you work in your jocks, do ye?' Ha bloody ha. I found that I actually felt guilty and over-compensated – putting in longer hours and working harder than normal. Imagine feeling fecking guilty? Ridiculous.

Anyway, back to that phone call from Mrs Kelly and our move to the rural idyll. The cottage in question was in the countryside near Dunmore East, County Waterford – my parents are both from Waterford and my mother and two of my sisters still live there. Mrs Kelly and I both love Dunmore, always have. It's the quaintest little place imaginable – a small, picturesque fishing village at the mouth of the river Suir that wouldn't be out of place in the South of France. We always vowed that we would live there some day, but saw it as something for the future – a retirement plan for our mid-sixties, perhaps, or maybe earlier if we had some money under our belts.

In the scheme of things, two people moving to Waterford is not such a big deal (especially when one of them is moving back to live near his mammy), but it was a massive decision for us to move our retirement plan forward by, oh, thirty years or so. We had spent so many years living and working in Dublin we were almost brainwashed into thinking that there was no way of earning a crust beyond the Pale. This seems completely ludicrous to me now, of course, but back then the decision to break our ties with the capital seemed monumental. Heroic. I don't recall there being any single thing that pushed us over the edge and made us go for it, but I do know that we got really sick of *getting through the week*. Putting in time. And for what, exactly? I suppose the endless slog would have been bearable if we had this wonderful life outside work, but we didn't even have that.

I knew the moment I walked into the cottage that I wanted to live there. I had a big goofy grin on my face going from room to room and Mrs Kelly had to give me a dig to play it cool in front of the estate agent because I was saying things like: 'God, that's lovely' and 'I love it, let's buy it.' The house was really quirky, to the point of being odd, but very homely. Plus, there was an acre of land with it. Mrs Kelly is a country girl at heart, having grown up on a farm, so the link with the land is more deep-rooted with her than it is for me. She likes having some muck and shit on her wellies, so to speak.

As for me? Well, let's just say I reckon there's a competent gardener somewhere deep inside me desperately trying to get out. When I moved to Dublin after college and had some disposable income for the first time, I went through a phase of thinking bonsai trees were the coolest things on the planet and within six months I killed off three of them that had a collective age of 450 years. Still, I had always been glad to hear my mother's yarn about my father sowing spuds in the garden of their first home, and how they'd kept some hens too that used to follow her down the road when she went for a walk with the pram. So, as I stood at the back door of the cottage that day, looking around the land, I could imagine myself growing vegetables – and growing my hair long and my beard longer, driving around in a battered old Land Rover with bales of straw in the back, wearing wellies and a waxed jacket, selling my prize cabbages at the local fête – you know, basically *The Good Life*.

In the end, we bought the house without having jobs or anything sorted out – that's the kind of cockiness the Celtic Tiger conferred on its cubs. Thankfully, Mrs Kelly got a job almost straight away with a manufacturing company in

Waterford city and when I mentioned the move at my office the company offered me the option of staying on to try and build up a business for them in the south east. At the time I was delighted (and, of course, it allowed me to put the tough choices on the long finger again). As I sourced an office to rent and furniture to buy, I imagined that the Waterford office would become my own little fiefdom with twenty or thirty people beavering away on my behalf while I drove around town in a big Merc throwing 'Regional Manager' business cards out the window and attending local functions as a captain of industry. I was in my element. We had taken the bold move and the sky hadn't fallen down.

The problem with bold moves, of course, is that they tend to generate a momentum all their own, like an errant snowball trundling down a slope. You try one hare-brained idea, it works, and all of a sudden you're more confident making another one. A few months after the move, and while I was still getting my bearings, Mrs Kelly decided she was going to jack in the accountancy and become a primary school teacher. She wasn't busy in her new job and all that sitting around practically idle from nine to five was driving her insane.

Two things happened around that time which gave us both a renewed sense of what's important in life. A very dear friend of ours became seriously ill with cancer and had to go through the horrors of chemotherapy to get himself better. Around the same time, a friend's sister died after a brief illness – she was a teacher and at the funeral the kids from her school did readings, sang songs, did the offertory procession. There was unbearable grief in the air at the tragic loss of a young life.

And yet there was also something else – a sense that this young woman had already had a more profound impact on her local community than most of us have in a lifetime. Mrs Kelly had always harboured a desire to be a teacher and her clock-watching in the job in Waterford suddenly seemed all the more futile. Strange that after all those years she spent studying, and having achieved 'success' in the modern sense of the word, having *arrived* – a good career, a nice house – she felt so unfulfilled. It goes to show that there are longings we have deep within us that money and *stuff* will never satisfy.

When she started her teacher training we went overnight from two very sizeable salaries to one, so we had our first mandatory introduction to downshifting. We were forced to change our thinking somewhat and not spend money on stupid things (like recording an album, for example), but apart from that we hardly noticed the fact that we had only one salary coming in. It's weird – it made us wonder what the hell we had been spending all that money on. Anyway, Mrs Kelly's passion for her new career seemed to put my own disaffection into sharper focus – it wasn't long before the sheen had worn off my new regional manager role and I realised I was still essentially a salesman, just with a bigger target. I hired one person, a young student from WIT, to do cold-calling and, God love her, it must have been the worst college placement in human history. This rather grim scenario – one college kid sitting in a big, draughty office with her boss – was as impressive as the personal empire got.

I was over at my mother's house for dinner one night and I was bitching about work and blaming my (late) father for having pushed me towards business studies in college as opposed to something funky like journalism or music. This was one of my favourite little tirades and one that I used frequently to justify why I was still stuck in a job I hated. 'Well, it's never too late,' said my mother matter-of-factly, laying it on the line as only mothers can. 'If you want to do journalism, go do it and stop talking about how hard done by you were.' At the time I was annoyed with her for depriving me of one of my great opt-out clauses, but she was right, of course. The thought festered for some time. What exactly *could* I do about it at this stage of my life, I wondered? The fact that Mrs Kelly was back studying gave me some wriggle room – I was the one earning the bacon, after all, I couldn't just give up, could I? On the other hand, the relentless, can-do attitude of the women in my life seemed to be silently judging me at every turn. One day at work, out of curiosity, I started looking on the web for journalism courses and was alarmed (and almost disappointed) to discover that I could actually do a part-time diploma down in University College Cork which would involve five hours of lectures each Saturday for a year. That sort of put it up to me, I guess – if I didn't do it, there was no one to blame but myself.

The course was essentially a creative writing course with all manner of other modules shoe-horned in. There was, for example, a financial accounting module which we called 'Advanced Payslip Analysis 101' because each week for a whole term, it seemed, the lecturer explained the contents of a payslip in great detail. But, in fairness, the course got me into the right frame of mind and because it cost a few bob, it forced me to start writing things and sending them off to newspapers. Shortly after I started the course I got my first break with *The Irish Times*. The editor of the Saturday magazine published a tiny feature about invisible keyboards, a technology that allows your mobile phone to project the image of a keyboard onto the table or desk in front of you. It was a 'blink and you miss it' start to my journalistic career – no more than 150 words. Nevertheless, it was a start. I was, I assured myself, an '*Irish Times* man'.

Later that month, they published a much bigger feature – a two-page, 1500-word piece, and I will always consider this my big break. I fretted and fussed over how my boss up at Head Office would take it when he opened the *Times* on Saturday and saw my big mug staring out at him. But I was really proud of myself. I got more of a buzz out of it than all the deals I made in ten years of selling combined – that's not an exaggeration. Friends and family were sending texts and e-mails. It felt great. A few weeks after that piece was published I got a payment slip in the post from *The Times* – and, armed with the comprehensive knowledge of payslips from the course, I could understand it too, which was nice. My first earnings as a writer. It wasn't bad money either, relatively speaking. At work in the draughty office later that week I started to daydream: if I could get X number of them published a week, I could give up my job.

One of the things which encouraged me to take the plunge and do just that was that I started to get quite a bit of work with a freelance writing agency. At the start, it was a couple of commissions each week which I could happily do without it interfering with the day job. Much. Around that time also, one of my weekly trawls through the recruitment pages in the paper finally threw up something interesting. There was an ad for a magazine called *The Gloss,* a new women's magazine looking for contributors. I contacted them immediately. Their first idea was for me to do a series on urban farming, a very wry look at the attempts of city folk to return to the land without getting their hands dirty, which, given our own move to Dunmore East, suited me perfectly. Most importantly, the editor commissioned me to write a monthly restaurant review. This was interesting, given that I profess to know very little about food, apart from the fact that I like to eat a lot. The fact that it was to be a regular monthly gig was really important to me – it meant a small measure of security. My first column. Yippee! During the meeting with the editor, we discussed rates and I was doing that old mental tot in my head – if I could get one or two articles per month into *The Gloss*, add that to the agency work and stuff with *The Irish Times* … hmmm.

Up to this point I hadn't let the writing impact on work too much, but I remember one day when I was in Head Office in Dublin I had to interview someone over the phone for a feature. It happened quite by accident, but I recall getting all hot and bothered and sweaty about the conflict of interests, scribbling down notes furiously in a notebook with one eye on the door in case anyone walked in. I started to think that the situation was ridiculous and that I needed to make a decision. At the time, Mrs Kelly was in the final stages of her teacher training – a three-week stint in a Gaeltacht. While she was away I had started to give some thought to giving up my job and the idea suddenly seemed to spiral out of control, with a mad momentum all its own. I welcomed Mrs Kelly home with the thought that she would put the brakes on this nonsense.

But, alarmingly, she didn't seem shocked at all when I gave her the news – if anything, she seemed determined to light a fire under my ass. 'If you hand in your notice next week,' she said, 'that means you would get your salary in August and maybe a commission cheque as well.' *Hand in my notice? Next week! Wait a fecking minute. When did we start talking about handing in people's notice?* 'You could put that aside as a month's wages for September,' she went on, 'by which time I will be back to work, hopefully.' *Jesus Christ, are we really discussing this?* 'What if you don't get a job in September?' I asked. 'We won't starve,' she shot back, quick as a flash. 'We'll manage. What's the worst that can happen? If it doesn't work out and we're on the breadline, you can go back to work in IT. It's no big deal. I've had my time off doing this course, so now it's your turn.' God, I love my wife.

Mrs Kelly has a wonderful knack of making a decision and then moving on to action immediately. I like to dither and procrastinate, to torture myself a little by analysing the living shit out of every single minute detail of each and every pro and con. Over the next few days I did exactly that. I knew, deep down, it was one of those moments that you should really listen to what your heart is telling you and try to drown out the objections from your head. And the head had quite a few. *You're throwing away ten years of work,* it warned ominously, sounding remarkably like my father. *Why did you bother going to university to do a business degree if you were going to piss the whole thing up against the wall in the end?* Then: *You're far too young to be opting out of your career – maybe in ten or fifteen years' time, I could understand, but not now. It's too early.* And: *Why don't you just forget all this nonsense and knuckle down? You're just a little de-motivated, that's all, but you can get it back.* And then my heart mounted a fight-back: *Don't listen to him! This is what you've always wanted. The opportunity you've always dreamed of is staring you in the face – stop fluting around and get on with it. Grab a hold of it – it's now or never.*

Back and forward like that it went in my mind *ad nauseam* and *ad infinitum.* The following week, I headed up to Dublin to meet my boss, deciding to decide on the way up in the car. If it felt like the right thing to do when I got there, then I would go ahead and do it. Even as I drove into Head Office I was still having doubts. It all seemed so final. Also, there were a few inexplicable, nostalgic thoughts creeping into my mind – ah this old car park: how many different cars have I parked in your leafy bosom over the years? These revolving doors that spin and spin and spin – how many times have you spun for me? A chat and some witty repartee with Trish on Reception – God, she's so nice. This old grey stairwell, that familiar musty smell, I shall miss you both. How many times I have climbed these stairs over the years? I remembered bringing Mrs Kelly up to see my desk during a summer barbecue. What if I actually miss the work? Jesus, maybe this is the wrong decision altogether. What if I miss the buzz of closing a deal? The witty camaraderie of sales meetings? *Feckin hell, do I really want to do this?* Oh, for God's sake, my heart told my head sternly, stop second-guessing yourself. You HATE sales meetings! *Just get in there and do this so we can get on with the rest of our life!*

My boss was just back from holidays. I remember he was tanned and relaxed-looking (not for long) and we chatted briefly about his holiday. I was actually shaking. I picked up a glass of water and had to put it down again to steady myself before I could get it to my mouth, my hands were shaking so hard. He clapped his hands together, saying something like, 'Have you got loads of deals to tell me about?' I swallowed hard and said: 'Actually no. I have something else to tell you about and you are not going to like it much.' I stopped and managed to take a drink of water. 'This is kind of tough,' I said (surprised at how tough it was). 'OK, OK, take your time,' he replied. He's a nice sort.

'I'm handing in my notice.'

And there it was. The magic words were out and once they are out, you can't pull them back in. My heart was pounding in my chest. My boss muttered something like, 'Ah, Mick, you're kidding.' I could see he was thinking: This is just what I f**king need. Two days back from holidays – I haven't even got through my e-mails yet.

I went on to explain the whole thing to him – that I was leaving to become a writer. 'Is there any point in me trying to persuade you to stay?' he asked. 'None at all,' I said, sounding a lot surer than I felt and wondering what would happen if he did try to persuade me. He asked me about the money side of it – would I be able to make a living? I told him, only half-joking, that I was voluntarily signing up to a lifetime of poverty.

When I had recovered my composure somewhat I went back to my desk and then told a couple of colleagues who were around that day. It's a nice moment when you can finally tell people you work with that you are leaving. I was on the receiving end of that for years and I always felt a slight twinge of jealousy. Mostly people were very supportive. But there was also something else in the air that I couldn't put my finger on but which has popped up a few times over the years since. Maybe suspicion. Distrust. Or envy? I'd like to think that hearing that someone is leaving to go off and follow their dreams challenges people to think about their own situation, especially if they aren't a hundred percent happy in their own work. Surprisingly, almost everyone I told mentioned some other thing they would like to be doing with their lives – one director told me he wanted to be a full-time photographer, another wanted to open a shop.

And that was pretty much it. It was over. I sat in the car and took a deep breath.

'Holy shit!'

I rang Mrs Kelly to tell her that I'd done the dirty deed. 'This is it,' she said, 'the first day of the rest of your life. Brilliant. Well done.' But I was churning up inside. What have I done? How the hell am I going to earn a living? What if I've made a big mistake? Motivated by sheer terror, I rang the editor of the *Irish Times* magazine whom I'd never actually met. I told her I was in town and asked would she be free to meet up for coffee. Thankfully, she said yes. We chatted for about half an hour and I explained to her that I was hoping to get more involved, having completed a journalism course and quit my job. I didn't tell her that I had quit my job an hour earlier, just in case I sounded a tad *needy*. That meeting was a nice way to take the edge off my apprehension about the future.

I had an odd mix of emotions in the weeks and months that followed. In one sense I was deliriously happy – I absolutely loved every minute of my new job and it felt like I was on holidays, permanently. The relentless complexity of my old job, the pressure of targets and meetings, the endless commuting – all these things quickly became a distant, unpleasant memory. Freelance writing has its own pressures, of course – you don't write, you don't eat – but they were my own pressures. And there's a purity and a simplicity to writing that is very sweet – you write a feature, it gets printed and then it disappears into oblivion. You move on to the next one. Your work is paced out in small, bite-sized chunks. I found that very refreshing.

But it wasn't all sweetness and light. As I've already explained, I am a heavy-duty worrier – if worrying was an Olympic sport I would podium-finish every four years. But even I was surprised at the amount of heavy-duty worrying I did when I left the job. I worried I wasn't busy enough. I worried that I wasn't working hard enough, that I wasn't earning enough, that I wasn't progressing enough or quickly enough. I worried about whether freelance writing was a viable career at all. At the start, I had maybe three or four hours of work on each day and so I couldn't shake the feeling that I had been laid off or that my new job was part-time and that I should really be out there looking for a supplementary source of income. Mrs Kelly didn't get a job until a week before the new term started in September so we had some anxious times, worrying about how we were going to cope if she didn't get a job. But thankfully, she duly got one in a country school in County Wexford, which took the pressure off somewhat from a financial perspective – at least it was a guaranteed salary every fortnight. Between the two of us, we were earning about as much as my old basic sales salary – that's pretty abysmal when you think about it.

I think most blokes are hard-wired to believe that they have a solemn duty to act as bread-winner-in-chief, and if we lose our jobs or suddenly find ourselves earning less than we used to, it can be emotionally emasculating. It's ridiculous, and incredibly tedious, really, that we feel this way because in most homes women have an equal role in bringing home the bacon – my missus had a better wage than I had before she left accountancy and that never bothered me. But still, I couldn't help feeling that I was doing something treasonous by giving up my job; that I was letting the entire male sex down. I was never idle – in my free time I did my best to earn a nomination for Housewife of the Year. I swept floors, dressed beds, did laundry. I indulged a dormant passion for cooking and did all the things in the kitchen that I always said I would like to do 'if only I had more time'. I made stocks and stews and soups and jams, patés, dressings and sauces. I baked hams and jointed chickens and marinaded meat. I made cookies, brownies, Madeira cakes, roulades, fruit loafs and tarts. I learned how to bake bread, so, even if I was no longer the primary bread-winner, at least I was the primary bread-maker. Now when I have a really shit day – where I seem to be writing with all the eloquence of a stone-age man making marks on a cave wall with a large rock – so long as I have taken a loaf of bread out of the oven I feel like I've achieved *something* memorable.

I went for walks with the dog, spent time in the garden; I read the paper over breakfast and if the weather was warm and the sun was shining I went to the beach for a swim. I did the crossword over lunch. I took up running too – something I figured I'd always hate but actually grew rather fond of. But while doing all these things, there was a gentle hum of guilt in the background. Looking back on that time now, I wish that I had cut myself some slack and enjoyed a well-earned rest from the stresses and strains of my old job and recognised that it was going to take time to build up my new career.

There's very little certainty in freelancing – when you have had a salary coming into your bank account on the same day each month for ten years, it's very hard to get used to payment coming in dribs and drabs. If you've had a bad week and haven't had anything published, no one is going to throw you a few bob just 'cos you're a nice bloke. You start each week with an empty payslip and work at filling it up. There's no health or life insurance, no more company cars, or company mobiles. There's no more reclaiming expenses. All the things that Jenny in Accounts used to look after are suddenly your own responsibility. You have to sort out your own tax affairs too, which, let me tell you, is very odd – the idea that I have to put some money aside (from the decidedly meagre amount that I get in) to pay a tax bill at the end of the year is a complete and utter mind-bend for someone as bad with money as me. In fairness, ex-accountant Mrs Kelly looks after the forms and calculations and all the nasty stuff for me. She actually looks like she enjoys it too: *Woo-hoo! A spreadsheet!*

For about six months, I busted a gut in a completely futile attempt to match my previous income. I took my old basic salary as a starting point and then divided that by the average payment I could expect per article. That left me with X articles that I had to have published each month – otherwise I would have to give myself a thousand lashes with a mighty whip and wear a cilice around my thigh for eternity. In some ways it's good to drive yourself towards some sort of a target, but in another sense I had to wake up to a fundamental reality – freelance journalists don't get paid as much as IT sales people. The only certainty in my new profession is that your income goes up and down, and there's very little you can do about it. That's just the way publishing works. You can have a bad week with nothing at all published followed by a great week when it feels like you're writing the entire newspaper. It's not your fault or to your credit either way – it just happens. I reckon that it took me nearly a year to chill the hell out and stop stressing about it. I suppose ten years of corporate conditioner takes time to wash out of your hair.

But the flipside. Oh my God, the flipside. The freedom of it. The flexibility. The sheer joy of working for yourself, being your own boss and working at something you love. Being constantly interested, challenged, proud, positive, excited, frightened. No more putting in time, trying to get through the week, living for the weekend. No more hating

Sundays and detesting Mondays. You would put up with all the uncertainty in the world for it. You would put up with lean week after lean week because you know now what it's like truly to enjoy work. You realise after a while that the security which a salaried job offers you is largely illusory and that financial security isn't all it's cracked up to be anyway. And here's the thing: when we were earning really good money we were discontented, agitated, annoyed, stressed, unfulfilled. And now we are happy. Go figure.

Chapter 3

THE COUNTRY SQUIRE

Moving to a leaky, rickety old cottage in the countryside sounds idyllic and in many ways it is, but at times the fact that it's leaky and rickety is just a pain in the ass and you have to stop yourself from dreaming of the bliss of a warm, dry, modern bungalow. Old cottages undoubtedly have oodles of personality, but it's a crabby old eccentric person's personality. You love that person a lot and greatly appreciate their eccentricities, but there are also days when you wish they'd just stop being so damned cranky. Gradually we've tried to put manners on this house of ours, but sometimes it seems to resent our very presence and throws up little quirks to test our resolve.

The original cottage was built about a hundred years ago and what was the old cottage is now one big sitting room. The previous owners added an extension to convert it into a three-bedroomed house. In keeping with the general oddness, the extension to the back looks like a Dutch barn, with the roof coming right down over the sides. There are lots of quirky little touches in the house that I like to think make up for some of its inadequacies (cold, damp, draughts, inhabitants getting The Consumption etc). There's one of those half-doors as the front door, for example – you know the type that you can open the top half or bottom half or both – and it immediately establishes the cottagey feel of the place. (That, and the fact that our house is called 'The Cottage' – I swear to God, there was a little plaque on the gate that said 'The Cottage'. I took it down because I thought it was really naff. Imagine living in a bungalow and calling it 'The Bungalow' or living in an estate house and calling it 'The Three-Bed Semi'.) When someone calls to the door we can open the top half – which doesn't exactly say, 'You're welcome', it says, 'You may be welcome, I'm not sure yet.' Handy.

There are some nice old salvaged stained-glass church windows around the place and a lovely beam going across the ceiling in the sitting room, which might well be keeping the whole house from falling down. There is a bizarre toilet under the stairs in the sitting room, but when guests come we usually send them to the upstairs toilet, otherwise they are basically taking a pee or worse in our sitting room while everyone else sits embarrassed on the couch, listening to noises that no person should ever have to listen to. There's a ship's porthole in the bathroom upstairs which, when we moved in, allowed you to look right into the bathroom from the landing – an interesting feature which must have made our first guests feel very uncomfortable indeed (we turned it into a mirror eventually). We have a Kilkenny marble fireplace in our bedroom and there are black flag-stone slabs on the kitchen floor.

Old houses like to have fun with their inhabitants – they throw *interesting* experiences your way and play with you, like a cat might play with a mouse before killing it. We came back from a weekend away late one Sunday night and the porch was flooded because the seal between the two halves of the door was worn away and, whatever way the wind was blowing that weekend, the rain had gone right in the door. Another biblical deluge caused a major leak in the roof and I went into the bathroom one morning to find water flowing down from the ceiling, down the walls and even dripping from the light bulb. When the elements come inside your home to such an alarming extent it's hard to feel anything other than completely and utterly useless. Up in the attic with my torch I could see the water cascading in from outside and the following day I discovered that the lead on the roof was cracked in two places; we had to have it repaired at some expense. We've been told that the whole roof needs to be replaced at some stage because the tiles are old and ill-fitting, but at the time of writing we haven't got around to it yet. For now, we lie awake on stormy nights listening to the soothing sound of the roof tiles clattering off each other in the wind – a gust gets in under one tile, then the next, then the next, all the way across the roof, like an elaborate Mexican wave. Most of the windows in the house will have to be taken out and repaired at some stage because they are in a bad way and very draughty. In the winter I stuff a sock along the seam of the window in our bedroom to try and halt the gale that blows in freely from outside. I did some amateur insulation around the windows and doors with that felt lining you can buy, but most of it is falling off now and blowing in the wind as a permanent indictment to how crap I am at DIY.

The house is really cold in the winter. In the summer it's just fair to middling cold. Builders in the old days didn't understand the restorative and heating power of the sun's rays and even if they did, they didn't have the glazing technology to take advantage of it, so, like most cottages, ours has really small, poky windows. As a result, very little light gets into the main part of the house and some of the rooms are liable to damp. We have a leather couch in the sitting room which gets a strange fungus on the back, I presume because there is damp in the floorboards. That's pretty unpleasant.

We've done some tinkering to try and improve the heat situation – we took out the fireplace and put in a solid fuel stove which we run almost constantly in the winter, with little care for our carbon footprint. I've discovered that you can bank down a fire in a stove and keep it lit all night long, which is one of the more useful skills I've learned. It means (environmentalists, look away now) you can theoretically light the fire in the autumn and keep it lit until spring. I doubled the amount of insulation in the attic and laid carpet in our bedroom to try and give us some warmth underfoot. We wrap up well – there are two rugs on our couch which get serious use and at night time in the winter we sit in two armchairs facing the stove with our feet as close as we can get to the flames without spontaneously combusting. There's a certain rustic charm to snuggling up nice and cosy on a cold winter's night but there's also a certain amount of just wishing the place was toasty and we could sit around naked. In my darker moments I dream of bulldozing the house and replacing it with a timber-framed, passive house.

My brother-in-law will kill me for putting this into print, but he swears that one night, when he was sleeping on the couch in our house, he was visited by a ghost in our sitting room. It apparently stood watching him while he lay para-lysed with fear. There was a good amount of drink taken over that weekend so there's a fair chance that the visitation was the result of our old friend, *delirium tremens*. But, to be honest, nothing would surprise me in our house. There was talk that an old man died in the house years ago, so perhaps it was him – but why he chose to haunt Mrs Kelly's brother in particular, leaving everyone else alone, remains somewhat of a mystery. So the house is leaky, rickety *and* haunted – well, that's just great.

Apart from spectral visitors, one of the other things that you have to get used to when you move to the countryside is that there is an unwritten rule granting little animals the right to come and go in your house with impunity, rather like a pack of marauding teenagers. Old houses just aren't sealed up the way modern homes are so there are literally entry and exit points all over the house – ancient, long established migratory routes under floors and up through old stone walls and drains. We moved in in August and as winter arrived the little critters finished their summer vacation in the great outdoors and came back inside *en masse* to the (relative) warmth. This first manifested itself in some classic moments in the bedroom (no, not *those* classic moments) – you know the ones, guys, where she wakes you up and says:

Girl: Did you hear that?
Guy: What?
Girl: That noise.
Guy: What noise?

You both listen intently. Cue rattling noise or clanking noise or the unmistakeable sound of little scratching feet in the attic.

Girl: That noise!
Guy: (Alarmed) Yeah, I hear it now, all right.
Girl: zzzzzzzzzzzzzzzzzzzzzz

Guy lies there with the whites of his eyes the only thing visible in the room. Girl slumbers gently, safe in the comfort that she has imparted her knowledge of the danger to the guy.

It's really unpleasant when you are lying in bed and there are things moving about in the attic, especially if you have only recently moved to the countryside. There's nothing like the combination of pitch darkness and little scurrying creatures overhead to bring on a palpable sweaty terror. You sit up in the bed and turn on the light. The noise subsides. You lie down again and switch off the light. A few minutes later it starts up again. The freaky thing was that this didn't sound like a mouse. A mouse scurrying around I could handle. But this was LOUD. It sounded like it could be a squirrel or a cat or something even bigger, like a herd of caribou. I lay there thinking the whole move to the country was a bad idea while the farmer's daughter slept blissfully in the bed beside me. I hate situations like that because there's absolutely nothing you can do about it – you're completely powerless. I mean, even if you could pluck up sufficient courage to get up into the attic in the middle of the night with a torch, what exactly are you going to do then? Chase whatever it is that's moving up there around the attic and then rip its head off with your teeth?

The next day, with considerable trepidation, I got up on the ladder, opened up the hatch to the attic and, with my heart in my mouth, shone a torch around to see if I could find anything. Nervously, I laid some mouse-traps. The traps didn't seem to make any discernible difference, so a few days later I got the big guns in. I can't understand how those pest control guys choose to do that job. While I had stayed on the ladder, poking my head into the attic tentatively, like the lily-livered city slicker I am, Rentokil guy clambered up and disappeared into the darkness without hesitation. He laid some bait, then came back a few days later and reported grimly that the little footprints in the bait led him to believe we were dealing with rats as opposed to mice. I wasn't surprised, but we were completely freaked out. 'Do you think your mother would mind if we moved in with her for a few days?' Mrs Kelly asked. Not at all, I said, packing up the car in a flash.

The poison killed the rat family off eventually, but a few weeks later I got a plumber in to investigate the fact that the cold tap on the bath wasn't working and after rooting around in the attic for a while he discovered some class of a dead animal in the pipe itself. 'Possibly a rat,' he said grimly, 'but don't tell the wife that.' No, I won't. 'Your water tank in the attic isn't covered over and when they're poisoned they look for water. That's probably what happened.' EEEWWW! I had to disinfect the water tank and cover it over with a plank of marine plywood to ensure *that* never happened again.

The Battle of the Attic flares up from time to time, usually when the weather is getting cold and just when we think we have snuffed out the last pockets of resistance. Lying in bed we hear something up there and I have to put down some more bait. It's anyone's guess how they get in – I've searched high and low for the smallest of holes and filled them all. One day I found a particularly large hole under the rafters and filled it with that expanding foam stuff (here's some free DIY advice – always use gloves when applying, if not it sets like cement on your hands and it takes a visit to casualty and three days of scrubbing to get it off). Anyway, the night-time noises don't bother me so much anymore, now that I am a dyed-in-the-wool country bumpkin and all. That said, I still don't like going up to the attic. It's a tiny, cramped little place, about a metre high that you can't even stand up in because our bedrooms have already annexed the space where the attic should be. I spent three of the most unpleasant hours of my life up there during our first winter in the house, putting down insulation, crawling around on my belly or lying on my back, dragging strips of insulation around behind me, coughing and spluttering and half-suspecting that any minute I was going to disturb a family of bears. I had nightmares for three consecutive nights afterwards and dreamt I had been buried alive. The attic doesn't feel like it belongs to us – I feel like I am trespassing any time I go up there.

I was in the kitchen one day when I heard what can only be described as a horrifying death-scream from upstairs. I ran up to find that Mrs Kelly had been rooting in the hot press for a towel and a bat had dropped down onto a sheet in front of her. As you can gather from the rat story, I'm not particularly brave when it comes to little critters, but I thought the bat was sort of cute, with a little scrunched-up face like Gizmo from *Gremlins*. My main problem with rats and mice is all the scampering and scurrying they seem to do (*Where is he off to next? Will he run here, will he run there? Who knows?*) – but this little fellow couldn't scamper at all. I'm sure bats are graceful animals when flying, but a bat trapped in a hot press can only crawl, and not very efficiently at that, with its wings getting all caught up in the towels and bed linen. I had him trapped in a cup in no time and brought him down to the end of the garden and released him near the compost heap, while Mrs Kelly still screamed her head off back in the house. He either (a) died a savage death at the hands of some wily predator, or (b) waited down there until night-time, then, under cover of darkness, flew right back to his nest above the hot press. We hope it's the latter, as I've since discovered it's illegal to remove a bat from your attic.

There was another scream one morning, this time in the sitting room, and when I went in, Mrs Kelly demanded, 'WHAT THE HELL IS THAT?' pointing at something on the ground. When I looked closer there was, indeed, a little animal sitting quite contentedly on the ground, staring into space. I thought it was a mouse at the time, but on reflection it may have been a shrew. We have one of those anti-rodent thingies that you plug in and it emits some sort of high-pitched noise audible only to rodents. Frankly, I don't have much faith in them because one day I was watching telly and I saw a mouse run right past it (then stop, give it two fingers!) and then disappear into the floorboards. But perhaps the little shrew was actually stunned by the device because he was frozen like a statue, so much so that I was able to put a glass over him, slip a piece of card in underneath it and bring him down to the end of the garden to keep the bat company. I reckon there is probably a little animal entrepreneur living down near the compost heap who has put his kids through college by selling maps of the garden to other animals that have gotten eviction orders from the house.

The cottage's unique eco-system doesn't stop at bats, rats and mice. It also plays host to a vast array of spiders. My sister lives in an old house too and she has the same problem – I guess it's that all the little holes around the place provide them with lots of places to hide out. These are not small, inconsequential spiders either – these are fecking tarantulas, extras from *Arachnophobia*, lurking with menace in the bath when you get in for a shower in the morning or scurrying across your pillow just before you retire for the night. Most mornings when I go into the bathroom there is an impressive complex of fresh cobwebs in the corners, engineered overnight. Spiders really are amazing creatures. When we moved in first, we used to vacuum them up a lot. Now we don't really bother – it's one of those jobs where you just can't stay ahead of the curve, so there's no point in trying. Also, the first year we moved in there was a veritable plague of flies upstairs – you would walk into a room and there could be a hundred flies sitting on the windows or walking upside-down on the ceiling, like Hitchcock's *The Birds* only with flies in it. God seems to have called off that particular plague, but I wonder what He has planned for us next? Locusts, perhaps?

When we lived in Gorey we did what we could to bring animals into the garden, though it's not easy in suburbia. We had a bird table with seed and nuts, but the birds stayed away and the nuts would rot after a few weeks. In the end I just gave up. If I put nuts out here, about fifty birds arrive within minutes, perching on every vantage point in the hedges and trees nearby before swooping in and half-killing each other to get at the grub. After buying nuts by the tonne for a few weeks I gave up trying to satiate their voracious appetites. Rather than trying in vain to attract animals, in the countryside it's a constant battle to try and keep them out, or at least to put some manners on them. You gotta love that about nature – it's not polite or courteous. It arrives and then takes over.

When we bought the house, the previous owner offered me the contents of the garage for €500 – the contents included the tractor-mower, a huge big industrial-sized strimmer (so that I can stand around staring at the overgrown ditches like a council worker) and more tools and gardening implements than you could shake a stick at. The mower is not in great nick and you have to sort of finesse it around the garden, but I don't really mind that. I am at my absolute happiest when I am out cutting the grass on that mower, though I can't really explain why. It's an arduous enough old job, even though I'm basically just sitting on my arse for two hours. In the summer I am only finished when I have to start again. But I don't mind. I suppose it's sort of a 'man out on his land' thing.

When the farmer who owns the fields around our house is out and about on his tractor, he gives a little beep of the horn if he sees me in the garden and I wave back – I like that little interaction we have. Sometimes I am on my tractor-mower cutting the grass and he is on his tractor and the moment has added meaning. I don't have a horn on my tractor, goddammnit, but if I did I would toot it to my heart's content. We are two country men hard at work on our tractors saluting each other – one with all the farming knowledge in the world at his fingertips, the other a complete novice thinking about things like: would it be cool to try drinking a beer while mowing the lawn? Strangely, the bonnet of my mower is adorned with stickers for the North American Hunting Organisation, the Chicago Motor Club and other red-neck organisations – I'm pretty sure if I ever did drink a beer while mowing, the guys at the Chicago Motor Club would be mighty proud of my efforts.

When we lived in Gorey I had a little wooden garden shed which I stuffed to the gills with equipment. It's not that I'm particularly great at DIY – I just love buying the equipment. Men are simple creatures and equipment completes us. I was glad to see, therefore, when we moved here that there is a massive garage, with row upon row of shelves and a large workbench where, if the mood takes me, I can drill things or saw things or stick things together or hack things apart. Living in the countryside allows you to indulge in all sorts of manly activities, if that kind of thing interests you. I'm a regular down the local co-op where I could hang out for hours, trying to work out what all the farming and DIY equipment is actually for. When I moved here I even bought an axe, the hunter-gatherer in me practically orgasmic at the thoughts of chopping wood for the winter. An axe is a great way to rid your body of any latent frustration – ten minutes chopping wood, I guarantee you, is more beneficial than an hour at the gym.

Last winter I discovered an old, rotten telegraph pole hidden in one of the ditches and when I realised that it would take me twenty years to chop it up with the axe, I took myself off to town to rent a chainsaw. When I went into the hire shop the guy behind the counter eyed my slight frame suspiciously. 'Have you used a chainsaw before?' he asked. I explained that I hadn't, but reckoned I'd be up to the job and I wondered was there any dirt in my fingernails I could show him that might convince him of that fact. 'I don't give them out to just anyone,' he said. 'They are really dangerous and if you injure yourself it will come back on us.' I assured him that I wasn't inclined towards litigiousness. It went back and forward like that for a while, with me increasingly feeling like this was some sort of bizarre test of my manhood. In the end I got all shirty with him and said, 'Well listen, I'm not going to beg you to do business with me,' preparing to leave. He was unimpressed by this show of impetuousness. 'Fair enough,' he said, 'but you should know that we're the only place in town that do 'em.' We were at an impasse.

Finally we agreed on a compromise whereby he would give me an electric model that was less powerful and had a cut-off mechanism so that I couldn't cut my feet off accidentally. 'It's a bit Fisher-Price,' I said glumly, staring with envy at the big petrol jobbie on the shelf behind him. When I got it home I switched it on and raised it up over my head like you would see the murderer do in a horror movie. That was fun (and probably explains why the guy in the shop was reluctant to rent it out to me). Then I attacked the telegraph pole with great gusto, admiring the slow growth in the pile of logs beside me and thinking of the long, cosy evenings in front of the hearth. Mrs Kelly pottered about nearby, pretending to be doing other things, but I know she was really either (a) attracted to the aura of masculinity that surrounded me while I worked or, more likely, (b) scared of her life I would do an amateur amputation on myself.

There's a rather odd thing about living in the countryside and it's this – a sense of isolation and a sense of community go hand in hand. In Gorey it used to drive us mad that we could hear our neighbours turning on their electric shower or plodding around upstairs on their wooden floors. In Dunmore the only noises we hear are cows bawling in the fields around our house – and, believe me, that's not as quaint as it sounds. I think Mrs Kelly really appreciates the cows – for her it conjures up 'home'. We go walking in the fields around the house with our dog and I think she can easily imagine that she is farming the land herself. In the summer months she leaves the window of our bedroom open and just after dawn I lie awake listening to birdsong. God, birdsong is irritating.

When we had hundreds of neighbours we actually felt more alone than we do living here with no neighbours at all. We live on a road that is about a kilometre long with just a handful of houses on it (for now). It takes time to adapt to that sort of isolation especially when you've been used to living in housing estates and apartments. I've always had neighbours, even when I was a young lad, so it's a little odd not to have houses on either side of you. Mrs Kelly didn't have neighbours when she was young so the feeling of detachment suits her – I like it too, don't get me wrong, but it does take some adjusting. I have a little routine before I go to bed, locking doors, putting on the alarm, that sort of thing. It makes me feel safer. Sometimes at night as I lie in bed and some noise outside or downstairs startles me, I wonder what exactly I would do if a burglar did come into the house. You can't exactly run next door for help. I suppose if we could overpower him, we could put him up in the attic and let the caribou nibble him to death.

Being happy in life is a multi-faceted thing, I've decided. It's far too simplified to isolate *one* thing, say your career or your private life or a nice house, for example, and then to claim that it's that one thing that makes you happy. But the sense of community and belonging which we feel here is certainly one of the major reasons we feel so contented. For all the reasons I've explained, that was completely missing when we lived in Gorey – and obviously I mean no disrespect whatsoever to Gorey when I say that. Of course a sense of community exists there too, but because we were never there, it didn't exist for us. Like many couples who buy in commuter towns that they have no connection with it, we had a sense that it was always going to be a temporary stop for us – a first step on the property ladder. From that perspective we were probably guilty of not trying very hard to integrate.

When you make a permanent home in a locality, a sense of belonging surely follows, especially if you are lucky enough to be able to make that home where you already have family and friends. In Dunmore, when we go up to the village we know people and they know us. Mrs Kelly says I over-do it, saluting everyone with great gusto like a Yank who has returned to the old sod looking for his long-lost relatives. She thinks I am overcompensating for the years in Dublin and Gorey when I knew no one. I just think it's nice to feel like you belong. After doing a year in the school in Wexford she got a job teaching in a local primary school which is all of a couple of minutes away in the car. I think it's nice for her to be such an integral part of the fabric of the community and, not surprisingly, she now knows about fifty times as many people as I do in the village. Sometimes when I pass by the school I can see her through the window doing her thing – I love that feeling that our professional and personal universe is so *compact*.

We were sitting outside a pub recently having a pre-dinner

drink and two kids that she teaches passed by and shouted out 'Don't drink too much, Miss!' You could see they were delighted at the idea of seeing their teacher outside her normal classroom role. We saw Mrs Kelly at the pub, they would be telling their buddies excitedly the next day, and she was drinking BEER!

Chapter 4

TUNNEL VISION

Only when the last tree is cut, only when the last river
is polluted, only when the last fish is caught,
will they realise that you can't eat money.

Native American proverb

It all started with a bulb of garlic. Had it not been for that one
measly, pesky little bulb of garlic, our garden would have
remained a nice, orderly affair with neat lawns and tidy flower
beds. Without the garlic, work in the garden would have been
about gently raking leaves from here to there, or lazy after-
noons spent pruning our prize roses. But no. Instead, the
garden was to become all about food production, and 'working in the garden' was to become ball-breaking, gut-busting,
hard labour. The bulb of garlic was our epiphany. Once we
knew its origin, there was no going back. It pushed us right
over the edge from relaxing hobby gardening to managing a
hard-core small-holding. Damn garlic.

Like most people, we're pretty conscientious about the
food we buy. We could stand in the supermarket aisles for
hours, agonising over a complex (and immensely tedious)
matrix of environmental and health issues pertaining to each

and every item of food we are about to buy: *On the one hand, this organic mango hasn't been sprayed with any nasties, which means it should be pretty good for me. But, on the other hand, it's been flown in from India, so it's an air-miles disaster zone. Maybe I'm better with a bag of Irish apples, but hold on, they're not organic so they've probably been sprayed half to death. What to do?*

Shortly after we moved to Waterford, we were in a supermarket one afternoon and Mrs Kelly was contemplating the pros and cons of buying a bulb of garlic. It didn't say on the shelf where it had come from and she actually went to the bother of seeking out a manager to ask him. That's not like her, incidentally – to be so, well, you know *anal* – but sometimes she gets the wind in her sails over something and there's no stopping her. 'It's from China,' said the manager, who had the look of a fella who had been roused from his tea-break and was in no mood for fielding questions about garlic from cranky customers. There was something about those three little words – *It's from China* – that seemed completely and utterly absurd to both of us. I mean, we're all pretty used to tomatoes from Holland and basil from Israel these days, knowing that we don't have the climate to provide an all-year-round supply of them ourselves. But garlic is one of the onion family. 'Is there not somewhere closer to home that you could buy it from?' asked Mrs Kelly. 'I'll have to come back to you on that,' said our manager friend, walking away, clearly concerned that she might lose the plot and go apeshit on him right there in the middle of the aisle. So, there we had it. Here we were, standing in a supermarket in the south east of

Ireland, on the western extremities of Europe, about to pay a measly 42 cent for a bulb of garlic that had been shipped a whopping five thousand miles from the farm of some undoubtedly poor farmer in China. This ludicrous scenario might be just about understandable if garlic *only* grew in Asia – but it grows perfectly well right here in Ireland. It just made no sense whatsoever.

The wellies 'n' wax-jacket utopian vision that had struck me the day the estate agent first showed us the garden was nice and all, but I never imagined it might actually come true. It hasn't done, not entirely at any rate, but if it does, I will continue to blame (or credit) the garlic. It's true that when we moved here first we both wanted to live in the country and have a big garden, but we certainly didn't see the land as a small-holding asset or anything like that. We weren't thinking: Ah-ha, here's an acre of land, we shall put it to work to feed us! I'm pretty sure we just felt excited that we had bought a house with a really nice, big garden and we were happy to tend to it like a regular garden – mow the lawn, weed the flower beds, mow the lawn, prune some trees, mow the lawn. That sort of thing.

Even if I had plans to become a small-holder, it would never have occurred to me that the new garden would be the place to execute those plans. To my mind the garden was a roomy, pretty, *al-fresco* dining location and not a lot more than that. The previous owners were keen gardeners and did some pretty impressive planting, so we just saw the garden as a nice place to spend time – it's surrounded by a hedgerow, which makes it good and private (and provides an interesting summer-vacation destination for the wildlife when they get tired of our attic). There are beautiful established trees in the front and back garden and we have planted more since we got here, including an oak tree which will be lovely in, oh, sixty to a hundred years' time. Out the back door of the house from the kitchen there's an enormous south-facing deck, which really is like another room. That's such a cliché, of course, but in this case it's really true. On lazy summer weekends we could spend entire days on the deck, cooking breakfast, lunch and dinner on the barbecue. So, in other words, when we moved here first the garden was about recreation.

This notion that we all have of our gardens being purely recreational, as opposed to functional, is largely a new thing, by the way. As recently as the early 1980s almost half the gardeners in Ireland grew vegetables in their garden. It is estimated now that the figure is lower than 10 percent. These days, gardens are considered to be an extra room and they are designed and decorated to continue the theme of the rest of the house. You can barely move at times with all the tables and chairs, inordinately complicated barbecues, scented candles, hot tubs, trampolines and deck heaters in the 'garden'. You only have to go to your local garden centre to know that this is the case – have a look at just how much of it is given over to selling traditional garden stuff, you know, like plants, shrubs and trees. (And around September each year, our local garden centre clears out most of the living things and morphs into a Christmas shop for three months. Quite what that has to do with gardening, I will never know. It's a shame, really, because their staff have fantastic gardening knowledge, but for those three months they put on Santa hats and provide the answers to far-reaching gardening questions like, 'What size batteries do these fairy lights need?')

Looking back on it with the benefit of hindsight, I can't imagine us living our lives now without producing our own food. It seems to complete the picture. But when we moved here first I was like Bull McCabe in *The Field* – I just wanted to get my hands on some land. I didn't have any great plans beyond that and certainly didn't think about the garden in terms of food production. In any case, while I was still working in sales, I didn't really have the time. I had my weekends free and I was quite happy to dedicate a couple of precious hours from my leisurely Saturday to driving about on my ride-on mower, admiring the view. But the little nudge from that Chinese garlic set the ball rolling on a more radical overhaul of our lifestyle. What started with Mrs Kelly's decision to plant a few bulbs of garlic in the garden to give two fingers to *the man*, spiralled completely out of control into an obsessive, highly addictive quest for kitchen-table self-sufficiency. This can be frustrating when you know as little about growing things as we do.

Anyway, as I've said, when Mrs Kelly gets her back up, she's pretty formidable and the little kerfuffle in the super-market aisle sent her into a tailspin of indignation about the evils of globalisation. The very next day we got to work on a vegetable plot down at the end of our new garden. We planted all manner of hardy vegetables – potatoes, celery, onions, leeks, carrots. And, of course, garlic. Lots and lots of garlic. Incidentally, this wasn't her first foray into vegetable production – in our little garden in Gorey she gew some lettuce on a tiny bit of reclaimed land behind the oil tank, which was a monument to her determination and a prequel to grander ambitions. But this new venture was about far more than a few heads of lettuce.

Ostensibly, the vegetable plot started as a protest about the lunacy of the modern food chain. It was about showing our friend in the supermarket that there are more sensible ways of procuring vegetables than shipping them half-way around the world. But as we planted those garlic cloves in some ill-prepared soil, something else happened too – a big piece of the jigsaw we hadn't really realised we were working on fell into place. It was as if planting the garlic was a sort of one-line personal mission statement, a comprehensive summing-up of what all the changes we had made were about. Living simpler lives? Check. Being thrifty? Check. Downsizing? Check. Getting back to basics? Check. Country living? Check. Environmental concerns? Check. Worries about food miles, food quality, slow food, organic food – check, check, check and double check. You have to admit, that's pretty impressive stuff for a little bulb of garlic.

Shortly after I left sales to start writing fulltime, I interviewed a woman called Rosie Boycott about her new book, *Our Farm*. By then, as I struggled to create a career for myself as a freelancer, I had plenty of free time on my hands and was spending a lot of it in the garden, up to my neck in muck. So from that perspective it was wonderful for me to meet someone who was so much further down the road to self-sufficiency than I was. Boycott was the first female editor of *The Independent* and the *Independent on Sunday* and edited the *Daily Express* until 2001. She was known as 'Rizla Rosie' in the UK press because of her campaign to legalise cannabis. She was an interesting woman to interview, though, disappointingly, she didn't seem to be stoned when I met her.

Boycott had set up a small-holding with her husband in Somerset and the book is a wonderful account of their attempts to make their new venture profitable. One of the things that she said to me during the interview about the sustainability of our food chain has stuck with me ever since. 'The entire food industry,' she said, 'is completely predicated on cheap oil. Not just oil, but *cheap* oil and that era is almost over. Our food is in constant transit, whizzing around on planes and trucks ... The saying goes that we are only ever nine meals from anarchy – in other words, if the supermarkets run out of food on a Monday, by Wednesday you would be willing to shoot someone else to get food for your kids. That's scary stuff. We all need to get prepared.'

Now, I know you're thinking: Oh come off it, Rosie, that's just 'end is nigh' stuff. But think about it this way: the fact that we in Ireland are an island nation, coupled with the pre-eminence of major international supermarket chains, leaves us particularly vulnerable to an oil crisis. What happens if it's suddenly economically non-viable to ship produce here because oil has become too expensive? We turn to local growers for replacement produce and we discover – shit, there are no local growers. *Hey, where did they all go?* Supermarket shelves are empty within days – and then what? But I have a loyalty card, we will say, banging on the door.

Our vegetable plot was not about surviving a supply-chain meltdown. Let's be honest: if such a thing happens, we're all f**ked and *our* little rows of vegetables won't help us much. Ok, so I might sit in the window upstairs with a shotgun, firing at starving looters trying to steal my carrots, but that will only buy me a few days at best. No, what our vegetable plot was really about was this: if Chinese garlic was symbolic of how complex and downright convoluted our world had become, then the vegetable plot seemed like a great way of restoring some balance, of getting back to basics, because producing your own food – whether it is some herbs on a windowsill or heavy-duty production – is one of the most basic, most *primordial* activities of all. We had acquired this considerable tract of land, and because of the time liberated courtesy of our job changes we had time to tend to it. The vegetable plot seemed to be a more logical way of doing things – a way of living a more straightforward existence and feeling better about ourselves and the world around us in the process.

Downsizing and growing your own food are two sides of the one coin. But the link between the two is not about economics – our vegetable plot doesn't save us a whole lot of money, though at the height of the summer when things are at full tilt you do get a nice surprise at the checkout. The link is mainly psychological. It's about a change of mindset. Think about our great, great, great ancestors – and I mean way back before industrialisation, before we knew our arses from our elbows. Back five hundred or a thousand or five thousand years. They lived their lives by a very simple mantra: grow a thing, eat a thing. They didn't think about *careers* the way we do. They didn't care much for job satisfaction or fulfilment. They didn't need to. They saw work as a means to an end. And that end was, purely and simply, having enough food. Contrast that to the complexity of modern life, to the layers and layers of noise we have built into our existence. For us, work is about work. It's an end in itself. Work is about producing money. And the proportion of our money we spend on food has become smaller and smaller. We spend more on mortgages, holidays, DVDs, home-cinema systems, clothes and all manner of clutter that we couldn't possibly need. And somehow, something tragic has happened. Having money to buy the clutter has become the driver for everything.

You do get a sense that society in general is on the verge of a major shift back to these basics, so now's the time to get involved while it's still cutting-edge. It's like we are emerging bleary-eyed after a protracted bad dream which has lasted almost half a century. Back before it started, pretty much every family in Ireland grew some of their own vegetables or kept a few hens or a pig, but, with the advent of the supermarket and kitchen appliances (most notably the freezer), we gave all that up. The 1970s, in particular, was the highpoint for an unshakeable belief in the abilities of science and technology to improve the quality of our lives. The iconic ad of that decade featured Martians laughing at the notion of us humans growing, digging, cooking and mashing our spuds. A generation abandoned the humble potato and switched over *en masse* to SMASH.

Growing your own food was for peasants. Convenience was where it was at. Why go to the hassle of rearing a chicken for the table when you can stick some ready-prepared chicken goujons into the freezer? Why go to the bother of growing peas when you have frozen ones? In the seventies, foreign holidays became the norm rather than the exception, so we jetted off to the sun and discovered pasta and fajitas. We took these new notions home with us, unwittingly playing our part in the dawn of globalisation. The ultimate outcome of which is, yes, you've guessed it – Chinese garlic! Yippee.

It's worth bearing in mind that when the whole world is a potential marketplace, the thing that matters most to commercial food producers is perishability. Not quality or taste or freshness, or localness, or sustainability, or seasonality, but perishability. The whole shooting match is set up around that one criterion – how can we make these products last longer so that we can ship them further, keep them on the shelves for longer and stop them rotting before they are bought? Everything that is wrong with our food chain stems from that – all the nasty shite they pump into food – preservatives, colourings, additives and the rest – are all used to prolong shelf-life. It doesn't take a genius to work out that when shelf-life is paramount, everything else takes a back seat. The very life force of the food which we need to survive is stripped out, all in deference to shelf-life.

The only way you can be a hundred percent guaranteed that the food you eat hasn't been messed with, re-packaged, altered, genetically modified, sprayed or riddled with additives, preservatives and colourings is to pluck it from the ground yourself. Six months or so after our supermarket epiphany, we got to pull our first bulb of garlic from the ground. We just presumed that our garlic would be the same as the Chinese garlic, only with less food miles. But there was an unexpected bonus: I stuck the spade gently into the ground and eased out the most enormous garlic bulb I'd ever seen. We couldn't believe it, it was the size of an orange, about three or four times the size of the ones you buy in the shop.

This garlic was the sweetest, juiciest thing I had ever tasted. When I bashed down on it with a knife (like Jamie does on the telly) juice oozed out onto the chopping board – how often can you say that about garlic? And what's even more breathtakingly amazing is that you can take one of those bulbs, take the ten cloves out and plant each of them in the ground and next year you have ten plants. It's that simple. It's simple to grow, needs no looking after, doesn't take up much space in the garden and tastes just a million times nicer that anything you will ever buy in a shop. So why IN THE NAME OF GOD are our supermarkets stocking cheap Chinese imports? I cannot tell you how much I enjoy the self-satisfied smugness that I feel when I walk right past those measly plum-sized bulbs of garlic on sale in the supermarket, looking down my nose at them. DAMN YOU TO HELL, cheap, nasty IMPORT! I don't need you! I'm garlic self-sufficient!

And this is a crucial point – the elation that you feel when you can walk past the nasty garlic imports is not lessened by the fact that you have to go ahead and buy other vegetables that you don't have the space to grow. It's probably true to say that most modern gardens won't allow you to become self-sufficient, but you can still dabble in a very serious way if you're so inclined.

One concept that has always stuck with me from Hugh Fearnley-Whittingstall's original *River Cottage Cookbook* was what he called the food acquisition continuum. The far left of the continuum is where you are completely reliant on super-markets for your food. The far right is complete self-sufficiency. Most of us will never reach the far right of that continuum in this day and age. No matter how much you pro-duce in your garden you will probably always have a hankering for the odd packet of Monster Munch or a Twix. But, as George W Bush would probably agree, any shift to the right is a good thing. This is a really important point because a lot of people might think, Oh I don't have enough land to produce everything I need myself, so why bother? But that would be to miss the point – any shift to the right strikes a blow against the mediocrity of the modern food chain. Even the smallest con-tribution from your own garden, whether it is a handful of basil for fresh pesto or some fruit for dessert, is a little victory in a very large war. And since the day we first started work on our vegetable plot, that's what we've been doing – winning little skirmishes and taking little steps towards self-sufficiency.

We were pleased to find out that growing your own vegetables doesn't require anywhere near as much land as you would think. I interviewed a guy one time who told me that small--holders in Portugal can make a living off as little as two acres and it is commonly accepted that it is possible to be completely self-sufficient on one acre of land. We have nearly an acre here, but only a very small part (maybe 10 percent max) is given over to food production – the rest is still, lamentably, dumb lawn.

A little later on, when we started to think about expanding our self-sufficiency drive to include the rearing of some animals, I approached the farmer who owns the land around our house about renting a small field. Quite why I would do this and ignore the 90 percent of our garden which was unused is anybody's guess – I suppose I was still in the 'garden as outdoor entertainment venue' mindset. Then one evening I was watching an episode of the iconic BBC show *The Good Life* – with yummy scrummy Felicity Kendal in her cutesy dungarees and crusty old Richard Briers. At one stage in that episode it showed how their garden was cordoned off into allotments for fruit and vegetables, chickens, two pigs (Pinky and Perky) and a goat called Geraldine. That got me thinking: why would I go about renting land when we have all this space here? Why are we so obsessed with lawns anyway? If the view from our back door was of a vegetable patch and a polytunnel in one corner, a few pigs in another and some hens scratching around on the lawn, would that be a bad thing? Would it be ugly? We reckoned it wouldn't. In fact, we figured it would be nice.

A vegetable garden will contribute more to your life than the lawn ever will. In fact, all the lawn does is suck up time, effort and resources. It gives nothing back, apart from looking pretty for a few days after it gets cut. But even on those days when it looks nice, you can't help thinking it's only a day or two away from needing to be cut again. In the summer months I get almost obsessive about the need to keep the grass cut short, clocking up hundreds of miles and countless hours on the mower. On top of that, I have to do a weekly sortie with the strimmer – I'm like the Grim Reaper on speed, wandering around the garden looking for errant blades of grass to chop down. That obsession comes with an environmental cost. The average lawn-mower produces as much pollution in one hour as forty cars, and my old heap of junk probably belches out more fumes than the M50 at rushhour.

Mrs Kelly taught me an invaluable lesson once (there were other lessons, I'm sure, but this one stood out). I was telling her how much I loved cherry blossom trees and thought we should get one to brighten up the garden. She scoffed and told me that cherry blossoms were city trees for city gardens. For the quintessential country feel, she reckoned we should opt for apple or pear trees, which produce equally impressive blooms but also produce crops of fruit in the autumn. I still reckon cherry blossoms would have been nice, but I can see her point. Regular flowers, bushes and trees are undeniably nice to look at, but they can't hold a candle to the ones that produce things you can eat. These are the true superheroes of the garden.

If you too are interested in producing your own food on any level, there are three books I would recommend that you get your hands on. I found them invaluable. The first is the *River Cottage Cookbook* which I mentioned earlier. The second is John Seymour's *The New Complete Book of Self-Sufficiency* which was first published in 1976 and really is the small-holder's bible. There are sections in that book that (sadly) most of us will probably never have any use for – skinning rabbits, felling trees, using a scythe etc – but they are fascinating to learn about all the same. The last one is Joy Larkcom's *Grow Your Own Vegetables.* There are many others. All of these books have a wealth of practical information to help you get that toe in the water, but, more importantly, they provide enough inspiration to help you get over your lack of ability. It worked for me.

It's a massive learning curve, of course, but I've found that blind enthusiasm can usually get you through the worst disasters. The main thing to remember is that most things *want* to grow and they will do so despite your best efforts to deprive them of that chance – planting them at the wrong time, wrong temperature, in the wrong soil, upside-down etc. It can be very daunting reading books and trying to absorb all the things that you have to do even before you start, and some of the books I mentioned above tend to assume a certain level of knowledge which you may not have. Don't let it bother you – if in doubt, just stick the fecking thing in the ground and hope for the best. Mrs Kelly is a great one for that. She just loves planting things to see if they will grow. I'm far more anal – I like having plans at the start of the year with diagrams and schematics. If she'd let me, I would probably have a list of garden goals and a family meeting at the end of the year to gauge progress.

Anyway, we plough along (forgive the pun). In the summer you really can't go too far wrong because the classic salad crops (tomatoes, lettuce, rocket, basil, radishes, cucumbers) are a cinch to grow. Lob every variety of lettuce leaf you can imagine into the soil and some of them are bound to come eventually. Radishes are also great if you want quick results – within a few days of planting you will have visible signs of growth and you will be able to harvest them within weeks. I think they make a great crunchy addition to a salad and they are also nice by themselves with some butter and lots of salt (and a glass of wine).

High summer is a time of year when you really see the potential for self-sufficiency, if not the reality. It's a time when your own produce starts to take centre-stage on the plate – it's no longer a little portion of a meal, it's the whole meal. If you want to have a more continuous, reliable and, dare I say it, year-round supply of vegetables and fruit, that's where things get tricky and you need some expertise. Behind any kitchen that has a year-round supply of vegetables from the garden is a meticulous, relentless planting and harvesting regime. It starts in January with the first, tentative sowing of seeds and it never finishes. We are nowhere even remotely close to having a year-round supply, but each year we do a little better. With gardening, you have to be patient, not just waiting for plants to grow, the weather to turn or the soil to improve, but also for your expertise to catch up with your ambition.

Our little box-room in the house becomes a mini nursery in early spring, with seed-trays snuggling up to each other on top of a warming mat and potatoes chitting away quietly on the window ledge. After a long winter, the first shoots of growth in a seed-tray really are tremendously exciting. For our first two years with the vegetable patch, we were like teenagers experimenting with sex and drugs for the first time – we tried to grow pretty much every vegetable we could think of. You name it, we tried it. After a few seasons of this wanton excess, we realised the folly of our ways – if you hate cabbage, for example, you aren't going to like it any more when it comes from your own garden. So what's the point in growing it? These days we tend to focus on the stuff that we actually like to eat.

Any old fool can manage to get a decent crop of any of the main vegetables. The real challenge is to give yourself a decent supply and to have the skills and space to be able to store what you don't immediately need. Take, for example, the really useful vegetables like onions, which are used in almost every meal. The absolute ideal would be to be able to produce enough onions so that you never needed to buy them at all. Now, wouldn't that be something? But to do that type of thing, not only do you need the gardening skills, you also need a complete change of culture. You need to go back to eating only what's in season, and in this day and age that's more difficult than we think. For example, are you going to give up eating fresh tomatoes from September to May when they're out of season or are you just going to buy the Dutch ones instead? If you're working to a recipe that demands, let's say, courgettes are you to going to abandon your plans to use the recipe because they are not in season? Are you going to stop eating oranges altogether because they can't be grown in our climate? Or do you feel that daily fresh orange juice is your birthright?

When we laid out our vegetable plot first it was a modest enough affair, tucked away neatly in the corner of the garden. Each year it annexes more and more of the lawn, which is a positive development to my mind. You plant six drills of potatoes and as you pull the last delicious spud from the ground you think: next year I will plant ten drills. Perhaps in five years' time I will have no lawn left at all and can consign the old tractor-mower and its redneck stickers to the scrapheap.

In our second summer in Dunmore, the vegetable plot was joined by a polytunnel which has turned out to be one of the shrewdest investments we've made in the garden. You've probably seen these in fields around the place as they are used by commercial growers for growing berries and salad leaves and the like. If you want to test your newfound 'I'm going to grow vegetables and I don't care how unsightly my garden looks' resolve, then polytunnels are just the job. They are undeniably ugly, sitting down at the bottom of the garden like a recently landed spacecraft, all plastic and makeshift wooden doors. But let us have no other badmouthing of the tunnel – we love it. It's our friend. It allows you to create a Mediterranean climate right here in Ireland, which means that as well as growing the stockpot vegetables you can get into more exotic fare – chillies, peppers, aubergines, watermelons and the like. Tunnels allow you to bend the rules of the seasons and extend the growing season at either end of the summer – very important in this Irish climate where our summer often consists of three moderately sunny days in August.

The alternative to a polytunnel is a fancy-looking green-house, but you get a lot more bang for your buck with a tunnel. Ours is 4m-5m wide and 10m long, and cost around €600. A greenhouse of those dimensions would cost ten times that. So think of a polytunnel as a sturdy, reliable work-horse, and a greenhouse as a bouffant show-pony. If you want an abundance of veggies and fruit for the table in difficult Irish weather, get a polytunnel. If you want your garden to be considered for a gong at the Chelsea Flower Show don't.

Polytunnels are difficult enough to erect. I was going to do it myself but, thankfully, Mrs Kelly quietly drafted in some family members to help, and her brother, in particular, seemed to know what he was doing. He immediately got stuck in, running builder's line around four pegs which marked out the corners, and he and Mrs Kelly discussed at length the application of Pythagoras's theorem to the problem of ensuring the frames were square. I tried to nod in all the right places and look interested (I didn't believe my maths' teacher when he tried to convince us there would be future practical applications for theorems). I kept to myself the fact that my plan (had I been left to my own devices) mainly involved digging.

Owning a polytunnel is very much like having a pet. It needs daily care. Because of the heat in the tunnel, the ground and everything planted in it dries out really quickly, so it needs daily watering. But we don't mind – these are labours of love. And, anyway, the tunnel is always a nice place to be. In colder weather it's usually slightly warmer in there and in the summer a sort of tropical intensity hits you like a brick wall as soon as you walk in the door. When it's raining outside, the tunnel lets you cheat the worst of the weather and work away regardless – the rain sounds even worse than it is, pelting off the plastic above your head, making you feel nice and snug. At the height of the summer and early autumn you can barely get in the door – it is resplendent with greenery, lush veggies and fleshy fruits.

The star performers of the tunnel are tomatoes, which thrive in the humidity and heat, and taste so different to the ones you buy in the shops. Plucked when ripe and eaten warm (as opposed to chilled, which is what we have become used to) they are sensational. We have become so used to the blandness of the ubiquitous supermarket tomato we have forgotten what a real tomato should taste like. An intense, sweet, luxuriant treat, not a dull, soggy, lifeless sandwich-filler. The plants get to almost two metres tall with big, thick stems and the pungent smell assails the nostrils. We typically grow little sweet cherry tomatoes, which are ready earlier, and the larger beef variety – my pride and joy that first summer season was a massive, red beef tomato that was as big as a grapefruit. I was reluctant to pick the thing, to be honest, thinking it would have been far better off in some competition, but in the end it gave its life to become a delectable bruschetta. RIP.

When you grow a quantity of tomatoes, or any vegetable, it's important to be able to deal with a glut. Tomatoes don't keep, so it's vital to be able to make use of the produce which you've slaved and fretted over for months. The internet came to our rescue and threw up lots of recipes – we made tubs of tomato purée for the freezer, which are dead handy for pasta sauces and pizzas. We also found a recipe for a tomato and courgette chutney, which enabled us to use up a prodigious glut of both vegetables. I'm no chutney fan, but I swear I was converted to it after tasting the one I made myself – I made about twenty jars in September of our first year and it was all gone by Christmas. Mrs Kelly even found a recipe for green tomato sour to use up the tomatoes that fell to the ground prematurely, and she reckoned it was nicer than my chutney. I told her it's not a competition (but my chutney was definitely better).

The cucumber plant is another voracious star of the tunnel. We use one plant only, but it produces enough cucumbers to last us from July to October. The skin is kind of hairy and needs a wash before eating to get it smooth the way they are in the shops. They taste great on their own or as part of cucumber pickle or tzatziki. They will grow to the length of the ones you see in a shop, in that sort of phallic shape – but we are never patient enough to let them grow that long. Courgettes are another easy one, indoors or out, and the plant gets enormous – it's like the *Day of the Triffids*. We have had courgettes up to half a metre long and, frankly, didn't really know what to do with them. We discovered that they are nice in a Bolognese sauce, on their own baked in a cheese sauce, or best of all simply cut off in thin strips and fried or barbecued with lots of olive oil and rock salt. Yum yum.

Peas and beans are another favourite. They are really unusual-looking plants, with narrow tendrils wrapping themselves around the support canes. I remember seeing a wigwam made of canes about three metres tall over in the west of Ireland once and there were runner beans growing up it – it looked spectacular. We tried to copy this, with mixed results. Eating a fresh pea is one of the great moments for any gardener. You just have to marvel at the pure genius behind a pod's design, protecting the delicate peas until they are ready to be eaten and then rotting away to nothing. It makes a mockery of modern plastic packaging.

We also made some good headway with herbs of all types and the tunnel is particularly helpful in that regard. Like most people, I was spending a small fortune buying herbs in the supermarket as demanded by recipes – basil, thyme, parsley, sage, marjoram, dill, rosemary. The little plastic cartons of herbs are really pricey and the contents are often uninspiring. They are undoubtedly handy, but not half as handy as having the herb growing down at the end of your garden. It's a doddle to produce a good crop of any of these herbs, and the ones that appreciate a little bit of heat love life in the tunnel. It's particularly pleasing to have a good crop of basil – I tried growing it in the kitchen lots of times, but was always frustrated at the meagre bounty and would nearly feel guilty about picking one or two leaves off the plant. Basil thrives in the tunnel and we get such a good crop that we can make big pots of pesto using massive handfuls of the stuff. That's more like it.

Like most Irish people, we love our spuds, so outside in the vegetable patch we have given over a lot of space to growing them. Again they are really easy to grow and don't seem offended by crap/unprepared soil, which ours really is. I still get excited every time I dig up a plant and find fresh spuds languishing in the soil. It really is like opening presents at Christmas. This year we didn't need to buy any spuds at all from May to September – next year I'm hoping we might be able to store some to keep us over the winter. We're a long way off being anywhere close to self-sufficient on the core group of vegetables, the really useful ones for stocks, soups and sauces: carrots, leeks, celery, onions. But we're getting there, slowly but surely.

We made some initial forays into fruit production too. Following the great cherry blossom versus fruit tree debate, we planted eight or nine fruit trees – apple, plum, pear and cherry down at the end of the garden.

(If you are interested in some native apple tree varieties, contact Seed Savers over in Clare – they are great guys and will give you lots of advice about how to plant and tend to your new purchase.) Our apple crop is, well, let's call it modest, but it's improving year on year. We were amazed at how well the plum tree has done in such a short space of time – who would have thought that plums grew so well in Ireland? I certainly didn't know that. I made plum jam to use them up. The cherry tree produces a lot of fruit, but this year the birds got to them before we had a chance to pick them. Let's not talk about the pear tree – it hasn't performed at all. But I have high hopes.

Outside the tunnel we have a small plot for fruit bushes: gooseberries, rhubarb (is rhubarb classed as a fruit? I can never remember), blackcurrants, redcurrants and raspberries. We had strawberries there too, but they took over the whole place so we took them up and moved them into a raised bed to put some manners on them. I love being able to wander outside in late summer and early autumn and grab some fruit for breakfast or dessert. You really can't beat it. I love raspberries in particular and the plants thrive – just when you think they are finished fruiting, a few more come along. The plant seems to know where to grow the little fruits to give them the best chance of making it to your mouth, in behind leaves and out of harm's way from birds.

Some of the best fruit we got, of course, we didn't grow at all – we have ditches around our garden which are full of brambles, and therefore blackberries. A landscape gardener, I am sure, would tell us to get rid of them and get a wall or a fence. But how could you cut down something that produces so much lovely grub? Sometimes you really just have to bow to Mother Nature's infinite wisdom. Think about it. As we approach late autumn and the first sniffles and colds of early winter arrive, what does she do? She piles hedgerows all around us full with more vitamin C then you could shake a stick at. You have to admit it, that's pretty impressive. Blackberry-picking, like all foraging, is an activity that can make you feel instantly connected with the past (there is fossil evidence that blackberries have been consumed in Ireland for up to 2,500 years) – far more connected with the past than heading down to the chemist's and paying €10 for vitamin C supplements, at any rate.

Given the amount of money we shell out on fruit and berries (I paid €8 for a punnet of blueberries recently), it's pretty ridiculous that each year millions of blackberries rot on ditches all over Ireland. If you take the time to pick them you can freeze them and enjoy them for months to come, or if you had some time on your hands you could try being a bit adventurous and making jam. It will take about an hour to pick enough of them to make a decent batch, but it's well worth it. Making jam is a complete cinch, compared to trickier ventures like making marmalade. I've tried marmalade twice with crap results both times – but my blackberry jam was a masterpiece, though I say so myself.

Every hour's work done in the vegetable garden has been repaid in spades (!) in the kitchen and on the plate. But it's not all about the culinary rewards. The time spent in the garden is pretty enjoyable too. I can't think of a better way to spend a day than mucked up to the eyeballs in the vegetable patch. In fact, even when I have a really tough, arduous day in the garden – where my back feels like it could give in and every joint and limb is screaming at me to stop – I still find myself at the end of it all wishing that the day wouldn't end. Imagine that? Without question, the most enjoyable, rewarding and, dare I say it, spiritual moments since we moved to Dunmore East have been spent in the garden, standing in the balmy air of the tunnel with a cup of coffee on a summer's morning surveying the progress, spending an hour with my back aching trying to get a fine tilth of soil for some new planting, seeing the first signs of life in a seed bed, weeding in the tunnel during a thunder storm, inhaling the heady aroma of tomatoes in high summer, digging up a magnificent purple kohl rabi and wondering what the hell to do with it, eating a meal where every ingredient has come from your own garden, and, of course, not forgetting digging up our sacred garlic – these have been some of the most honest moments of my whole life.

Chapter 5

HEN PARTY

If you get any semblance of joy from growing your own vegetables, I promise you, you will get serious additional pleasure
from keeping a few laying hens. Imagine having an animal that
lovingly produces an egg for you each and every day will make
you feel like a farmer without all the attendant crap that farmers have to put up with! Keeping hens for their eggs is one of
the most basic of human activities, and to get a feel for just
how fundamental it was to our forebears, just think about how
many modern expressions come from the world of poultry:
we talk about not 'counting our chickens before they hatch';
the 'pecking order'; we say someone is 'broody', or that our
troubles will 'come home to roost'; if someone gets above
themselves they are 'cock of the walk'; a man who is under the
thumb is 'hen-pecked' and therefore spends a lot of time
being 'cooped up', till eventually he looks like no 'spring
chicken'. You get my drift.

I was on a cookery course recently and someone asked the
chef where she sourced her eggs – a debate ensued about the
difficulties of getting decent eggs and I sat there wondering
should I tell everyone just how easy it is to have your own
supply fresh each morning. Keeping hens is something that is
very accessible to the vast majority of us. Keeping cows and

goats and pigs is probably a bridge too far for a lot of modern gardens and our modern lifestyles, but hens are different.

There are very few gardens that are too small to support a couple of laying hens, and I mean very few. We didn't know it at the time, but we could easily have had hens in our little garden in the estate in Gorey, no problem at all. Admittedly, it comes back to what you think your garden is for – if you leave hens out and about in your garden they will mess things up a little on you. It won't be anything serious, but there will be quite a lot of pooh around the place and they like taking dust baths in your flower-beds. But the payback! You will not believe how good those eggs are. I also think there's nothing nicer than seeing a few hens wandering around on your lawn. Think about it this way – what does your pet cat or dog really contribute to your family when the chips are down? Lots of love and laughter and companionship, I hear you say. Big fecking deal! You can't pay the rent with love and laughter. At least a hen will pay its way by producing some eggs.

Let's talk about the eggs we buy first of all and try and put some perspective on the situation. As we know, food producers are bastards and they are constantly trying to sell us a pup when it comes to the food they're trying to get us to eat. If you pick up some eggs in the supermarket and they are labelled any of the following: 'Farm Fresh', 'Country Fresh' or 'Naturally Fresh', you would be entitled, would you not, to have a fairly romantic image of the environment the hens that laid those eggs are reared in? Most probably you would be imagining a nice, spacious farm in the countryside? In reality the image you have is likely to be entirely wrong, which shows just how misleading that much maligned word 'fresh' can actually be. Pretty much anybody can use the term 'fresh' when it comes to selling eggs and not be remotely concerned with the conditions that the hens are kept in. If you could bottle a fart and sell it, you would be perfectly entitled to slap a label on the bottle which said 'fresh fart' – I mean, clearly the air inside the bottle wouldn't be fresh at all, but no one could pull you up on the label because it means *nada*, zilch, nothing.

So, what that means is that the hens that produced those 'fresh' eggs could be reared in what you and I would consider quite appalling conditions. These conditions have arisen presumably not out of any penchant for sadism on the producer's part, but rather out of the desire to maximise the utilisation of space. In other words, if a producer has a large barn, he wants to cram it to the gills with laying hens so he can get as many eggs as possible from them. After all, eggs are pretty cheap and you need to sell a shed load of them to make money. Incidentally, by expecting cheaper and cheaper food and always voting with our wallets, all consumers (that's you and I) play our part in bringing about these circumstances. It would be easy to pretend that these producers are just unscrupulous evil folk, but deep down we know that it's far more complex than that.

If you're buying eggs that are not free-range, you're buying eggs which were laid by a caged hen. Up to 70 percent of laying hens in Ireland are caged. In a battery-rearing operation, four hens typically share a 50cm-square wire cage. It might be useful to take a moment to imagine just how small 50cm is – that ruler that you used to have in school was about 30cm long, so the cage is roughly two of those. Unfortunately, the hens themselves have a wingspan of about 80cm, which is quite a bit wider than the cage, so they will never have the option of giving the odd flap should the mood take them. And don't forget there are four of them in the cage anyway, not just the one. The cages are then stacked on top of each other, again to 'maximise space utilisation' – so the poor old hens are quite literally sitting on top of each other. It's a miserable existence. But perhaps they can be thankful for small mercies, like the fact that they weren't born male. Males are obviously useless when it comes to laying eggs and since they are not the right breed to be used for their meat, they are killed as day-old chicks by gassing or by a machine that minces them (presumably alive). Over one million male chicks are killed in Ireland each year.

Intensively reared hens are kept in rooms which are deliberately kept free of natural sunlight because natural sunlight has the annoying habit of giving way to darkness at night time. Like most animals, hens like to rest when it's dark and, while resting, they are not laying. So, large-scale egg producers like to use artificial light instead, which keeps it looking like daylight pretty much all the time. Hens are not the brightest animals on the planet and because they think it's daytime, they are fooled into laying more than their standard one egg a day. This is good news for the producer (vastly improved profits and all that) but it's terrible news for the hen, and producing such a large number of eggs with inadequate resting time means one thing – an early appointment with a hatchet and a final resting place as part of a chicken stock cube.

Much like ourselves, hens need sunlight to stock up on vitamin D, so it's not surprising that when you keep them indoors in these conditions disease is rampant. This means that producers have to lace their food with antibiotics to keep disease down and then with an artificial yolk colouring (called xanthophyll) to return the eggs to their proper colour. They do something similar in Augusta each year for golf's US Masters – they spray the fairways with a nice dark green dye to make them look more impressive for the TV cameras – only a fool would think that the grass is in better nick as a result. When you think about it like this, those 'fresh' eggs don't sound very 'fresh' at all.

Any group of hens, even free-range ones, will establish a detailed pecking order the way a street gang might, relegating some of their number to the lower castes of society while others sit atop. They enforce this pecking order mercilessly and quite viciously at times. It can seem kind of cruel when you introduce a new hen to a flock and all the others peck the living daylights out of it for a few days, but that's how hens organise themselves. After a few days it stops, once the flock's new status quo has been established and each hen knows its place. When hens are kept in cages they turn into utter assholes (as you would) and just peck each other for something to do, pulling out each other's feathers and causing injuries. Some battery-reared hens end up almost entirely bald, and many die from their injuries.

When we got our hens first I was absolutely blown away by the eggs. They tasted and looked different to any egg I'd ever eaten. For starters, the yolks were a deep orange colour rather than yellow, which was a surprise – they were so different in colour that at first I wondered was there something wrong with them. I gave six eggs to my sister and she told me later that she actually considered dumping them because they looked different to anything she'd ever seen. I wondered was I just looking at our eggs through rose-coloured glasses, so I did an experiment (I know, I know, nerd alert!) in which I cracked one of our eggs into a glass and put it alongside another glass which had a free-range, shop-bought egg in it. The difference in colour was striking. The free-range egg was a dull, insipid yellowy-grey colour, while ours was the aforementioned brilliant orange. It's the same when you cook them. You know, when you are in a hotel or the airport getting your fry-up and you lift the lid on the scrambled eggs and it's this big mass of dull, yellowy-grey mush? Well, when you make scrambled egg with your own eggs it looks completely different – it's a deep, vivid orangey-yellow colour. Once you've tried the latter, of course, you can never go back and generally you find yourself avoiding eggs altogether when you stay in a hotel.

But let's be clear – we don't do anything particularly special with our hens. They have unlimited access to fresh air, fresh grass and we feed them an organic feed, but apart from that there is no special care given to them. So how could the eggs they produce be so much better and so much tastier than free-range ones?

The main EU regulations which are forced on free-range egg producers are as follows: (a) the hens have to have continuous daytime access to open-air runs and (b) there should be no more than 1,000 hens per hectare. A hectare is 10,000m², which means that each hen theoretically should have access to 10m² of space. That sounds reasonable enough, doesn't it? My little cubicle back at Head Office was a lot smaller than that. So, clearly, that doesn't explain why the eggs are so crap. One suggestion is that the problem with free-range eggs comes from the regulations about exits to the open-air. Currently the legislation says that there must be one exit of 18 inches (45cm) in height and 24 inches (60cm) wide for every 250 hens – the argument goes that when hens hang around these little exits (which they tend to do as they are naturally cautious animals) they quickly become congested and the majority of the hens get more or less stuck inside.

It may also come back to what the animals are being fed. The free-range regulations do not provide any guidance on what the hens are to be fed, so maybe that's the problem. The regulations on organic eggs, on the other hand, have quite a bit to say about what the hens can be fed – an organic feed, and the grass that they are grazing on must also be organic. There are studies which indicate that organic eggs are richer in omega 3 and have significantly higher amounts of vitamin A, vitamin E and beta carotene than non-organic ones. But I've had organic eggs from the shop and they still don't taste as nice as ours (maybe I am just biased). Perhaps it is because commercial hens are fed exclusively on grains while the ones you keep yourself get the odd bit of leftover fruit and veg. Hens like a bit of variety in their diet. The only conclusion I can come to – and it is a rather depressing one – is that large-scale commercial egg production, regardless of the condition the birds are kept in, just cannot provide the high-quality eggs you get from a small, home-reared flock.

Anyway, the point that I am trying to make is this. If you are buying eggs in a supermarket, remember that many labels don't guarantee much of anything, but organic is better than free range and free range is infinitely better than 'fresh'. Better still, try to buy your eggs from someone who has a relatively small number of hens – try your local farmer's market or find someone locally who is keeping a few hens themselves and might be glad of the few extra bob or even some class of a barter arrangement. Don't worry if they don't have the free-range or organic designation – you might get spectacular eggs from a farmer down the road who wouldn't qualify for organic status in a million years. If you want the BEST eggs of all, get yourself a couple of laying hens. I promise you, it will be worth it.

We got out first batch of hens (four Rhode Island Reds) in our first summer in Dunmore. I've been asked a few times by people who are interested in getting some hens: Where exactly do you get them from? It's a good question. The mother-in-law sourced ours for us. She seems to have a dealer for these things much like some people have a dealer who can get them drugs when they need them, no questions asked. She made a few calls and a few days later some hens arrived in the back of her car, huddled in a cardboard box. If you don't have a mother-in-law/dealer like I do, I would recommend that you check the classified ads in *Buy and Sell*, *The Farmer's Journal*, local newspapers etc. There's always someone selling a couple of hens. There is also a national organisation, the Irish Poultry Fanciers Association (I always think the name sounds a bit fetishist) which hosts annual shows where you can go along, learn a lot about hens from true experts and buy some hens too, including rare-breed varieties.

Let's talk a little bit about the downside to keeping hens just so I can't be accused of painting an overly romantic picture of the thing. Hens are fundamentally high-maintenance animals without being particularly high-maintenance to look after. That probably sounds like an oxymoron, so let me explain. Bottom line, you will have to feed them once or twice a day and you will probably need to lock them in at night. And that's pretty much it – which doesn't sound too bad, does it? But beneath that civilised veneer there are always little dramas going on. We currently have about ten hens and it really is like a daily soap opera or dealing with a bunch of hormonal teenagers. There are scuffles over who gets into the nest to lay an egg first. There are huffs and squabbles, replete with noisy squawking. Someone stops laying. Then starts again. Then someone else stops laying. There are noisy changes in the pecking order. They start laying in the ditch instead of in their nesting box. They fly out over the fence that you spent five hours putting up to keep them enclosed. Or, worst of all, a fox gets them. Mrs Kelly reckons that I am obsessed with our hens and she is probably right – outside of my basic responsibilities to keep them fed, I do seem to spend an inordinate amount of time fretting over them, but in my defence I just LOVE eggs.

Then, hens shit everywhere. They seem to be able to produce amazingly large amounts of it. They shit on the lawn. On the deck. On the driveway. In their house. On their house. On a windowsill. On a path. You name it, they shit on it. At nighttime, they shit while asleep on their perch, so about once a fortnight one or other of us has to clean out their house, which is an especially unpleasant experience – mucking out dung-covered sawdust and straw with a putrid, gut-wrenching stench assailing your nostrils. People will tell you that hen shit is great for the compost heap, as if you're supposed to jump up and down with excitement about the potential for great compost and forget about the regular unpleasantness of cleaning out their house. You won't. It's a terrible job and no amount of compost will make up for it.

You can go away and leave your vegetable patch for a few days and while some plants might die off, there won't be any longterm damage. Hens, on other hand, tie you down. We think it's worth it, but clearly that doesn't suit everyone. I've talked to other people who keep hens and some of them have used feeders (or hoppers, as they are also called) with great success – the idea being that you pour in a load of feed but it only lets out a small bit at a time so, in theory, you can feck off for a few days and they will be sorted for food. We've never been a fan of them because I think they attract rats and they often get clogged up when it rains and food gets a little wet. Owning hens also means living in mortal fear of the fox. We've had two or three visits since we started rearing hens and I can tell you it's really unpleasant if a fox gets one of your hens – don't get me wrong, we don't start crying or wearing black when it happens, but it's not a very nice experience. The upshot of all that is that if you're going to be going away, you'll probably need to arrange for someone to come in and check up on them.

You don't need to be too fancy with their housing. Hens have pretty simple requirements when it comes to that. The place should be secure and dry. It needs a roost or perch for them to sit on at night and a nest for them to lay in. And you have to be able to get at the eggs. And that's pretty much it. There's a company called Omlet that sell a purpose-built, small plastic hen-house called an Eglu, which has a small run attached – perfect for the first-time hen owner (it houses about four hens) and people with small gardens. It's light, so you can move it around to fresh grass every few days so that they don't make a mess of your lawn. There are also more elaborate wooden arks available, but I found these to be excessively expensive and, cheap fecker that I am, I decided to build a hen-house myself. I found a detailed design for an ark-like coop on the web and then completely ignored it, feeling that it would be better to follow my instincts. I eventually used so much wood in the construction that the owner of my local DIY shop was able to hire some extra staff and pay off his mortgage early. Now, whenever he sees me coming he praises God and shouts something in to his wife in the back about how they will 'eat well tonight'.

That first house that I built (yes, there have been others) was a triangular shape, about a metre and a half long and a metre tall, like a really small two-man tent. One side of it opened out on a hinge to allow you to get at the eggs. It had a perch inside – basically a small, horizontal bar, 10cm or so off the ground, where the hens sat at night, discussing how crap the house was. I spent about three hours building a small ladder for the hens to climb up to the perch, using copious quantities of wood glue and equal measures of sheer grit and determination. My friends laughed a lot at the house, but they reserved special mirth for the ladder.

Hens like to have a secluded spot to lay their eggs in – it's their natural instinct to lay in private. A caged hen will actually hold onto her egg as long as possible because she doesn't have the privacy her instinct tells her she needs; you can imagine that holding onto an egg is not good for a hen. Then, if you leave your hens out in the garden they may take to laying in a hedge or ditch, which sounds kind of cute in a sort of 'let's go foraging' kind of way, but actually is a pain in the ass because invariably you won't be able to find them. We had a springer spaniel called Ozzie who used to do a few laps of the garden in the mornings to see if he could find any eggs, so it was a race against the clock and a battle of wits as to which of us would find them first. One day I found him with his head in the ditch and his arse in the air, tail wagging furiously, and I just managed to pull him out in time – he had located five eggs and was busy slurping his way through the first couple.

You can discourage this behaviour in your hens by providing them with somewhere cosy in their house to nest. A small cardboard or wooden box with some straw in it is ideal and it is from here that you will be triumphantly plucking fresh eggs each morning. We've been keeping hens for nearly three years now and I still get a warm, fuzzy feeling when I see a couple of fresh eggs waiting on the straw for me. Perhaps I'm a little weird. If you have more than three or four hens, it's a good idea to provide more than one nesting box, otherwise there will be competition among the hens as to who gets in first.

To give you an idea how utterly infuriating hens can be: we have two nesting boxes in one of our houses and for some reason they all seem to like laying in one of them but not the other. Some mornings I go out and three of them could be cramped into one box. Only one of them is laying; the other two are trying to secure their place in the queue and 'encouraging' the laying hen to speed things up a bit by sitting on top of her. Meanwhile, the other nesting box lies idle beside them. God, they're painful.

I attached a three-metre run made of chicken wire to the first house. When we got the hens first we were wary of our dogs and generally hadn't a clue what we were at, so we initially kept them in the run all the time. Usually hen arks are designed so that the run can be moved around to fresh grass, but because I didn't follow the design, our one weighs about as much as a small car so it can't be lifted at all – this means the grass in the run gets manky in a few days, all matted with hen shit and trampled down by little claws. It wasn't long before we started to leave them out, at first supervised for an hour here and there, and then, eventually, we left them out all the time. It's a nice sight seeing them having the run of the garden. Hens hate the dark so they always return to their house when the light fades. That means, in theory, there is no 'rounding-up' to do in the evenings, a disappointment for the wannabe farmer. But you can never be sure with hens, so you do have to check that they are in at night.

We feed them a mixture of rolled barley and organic layer's pellets, which we get from Morrin's in Baltinglass. It's expensive, but I buy about three or four bags at a time and it keeps us going for about four months. I'm not sure how much better it is than a regular layer's pellet, but at least we know there's no chemical shite in it and I guess if we ever thought about selling a few eggs at a market, they could be certified organic. Any rotten fruit or cooked vegetables from the kitchen go down a treat with the hens. They love kiwi and banana, mashed potato (especially if it's warm), and anchovies are a particular favourite. When I make a juice I give them the pulp from inside the machine – I throw it in their bowl and then stand back while they kill each other trying to get at the spoils.

When you get hens at first they are usually at what is called 'point of lay', which means they are about seventeen weeks old and are about to start laying. It's important to point out that you don't need a cockerel in the flock for your hens to lay eggs – a hen will lay eggs regardless, but the eggs aren't fertilised when there's no cockerel on the scene, so they will never hatch into chicks. When we got our first batch of hens, I would practically run out each morning to check on progress and return egg-less and disappointed. Mrs Kelly put a yellow table-tennis ball sitting on the hay one morning and it fooled me for a couple of seconds. How she laughed. When the first egg finally came, it was indeed a moment to savour. The first three or four eggs were a little small, but after a week or so all four hens were laying and the eggs were as big as those you would buy.

The number of eggs that we get varies from day to day. Our first four hens were mighty regular and used to lay an egg each most days. But now that we have a mix of old and young hens, there is more variation. Some days we get five or six, other days we might only get three. Sometimes you get an enormous dual-yolk egg, which I love because you tell yourself you are having one egg, but really it's two.

We eat lots of eggs. Boiled, poached, scrambled. Omelettes. They look and taste spectacular. I promise you, you will never see anything like an omelette made from your own eggs – it's a big glow of orange on your plate. We also use lots of eggs in baking and cooking, and we give away the balance. Last summer I had a bit of a barter arrangement going with a neighbour who gave me fresh fish and some sweet apples in exchange for a good supply of eggs.

We get over thirty eggs a week so I expect I will keel over with clogged arteries any day now – either that or all the fretting over the hens will cause an aneurysm. Either way, those feckers are putting me in an early grave. For a long time we were told that eating too many eggs is bad for you. Nowadays we are being told that it's ok to have an egg a day – they even have a slogan: 'An egg a day is OK.' (I wonder how much money they paid the ad agency to come up with that one?) I think the so-called experts are often very confused about what is good for you and what's not, and the most sensible approach is to do your own thing. Mrs Kelly's grandfather had a boiled egg every morning of his life and he lived to see his nineties. I know that's circumstantial evidence along the lines of a smoker who says, 'I had an uncle who smoked a hundred fags a day and lived until he was eighty', but, seriously, how can something as natural as an egg be bad for you?

Hens are pretty sociable animals and have a clearly defined order to their day. Morning times are for eating and laying eggs. In the afternoon they like to wander around together, foraging and scratching in the grass for worms, slugs, spiders, seeds. Hens have amazing eyesight – one of the funniest sights ever is a hen running after a fly or insect; they almost always catch them too. When our first batch arrived, I was surprised at how attractive they looked. I had assumed they'd be kind of scrawny, ugly things, but in fact they are quite proud and aristocratic-looking. They have a shock of rusty feathers and keep themselves very clean. They take regular dust baths to maintain the quality of their feathers. It's a strange sight to behold if you don't know what they are up to (which I didn't at first). Basically they scratch a hole in the dirt and then lie in it, pulling dust into their feathers with their claws. They will lie like that for ages and at the end stand up and shake themselves, sending a vast cloud of dust into the air, like a small nuclear explosion. Battery hens will try to give themselves a dust bath on the cage floor, purely because it is their instinct to do so – you can only imagine how that works out for them. Most of the time, hens are very quiet animals to have around. When hens are contented they let out the occasional *Bok bok*, much like a cat would purr. The loudest they get is when laying an egg, but then if you had to push something that size out of you each morning, you'd be loud too. It starts as a regular, incessant *Bok bok bok*, culminating in a noisy climax of *Bok bok bok bok b-caw*!!

Hens love flapping their wings and breaking into a run, which makes you feel even worse for their battery reared buddies. The first thing they will do when let out of their house in the morning is run off, flapping their wings furiously; it's the equivalent of you stretching when you get out of bed in the morning. That still makes me smile. You can really understand where they got the idea for the movie *Chicken Run* when you see them running off like that in a line – they look like a squadron of old fighter planes taking off on a bombing raid. Hens look really stupid when they run. It's as if running draws attention to the fact that they don't have arms. Try to imagine what you would look like if you ran with your arms pinned to your sides – that's what a running hen looks like. A battery hen doesn't get any exercise at all and so suffers from skeletal and muscular weaknesses.

TRADING PACES

The hens keep our dogs amused. We were a little worried when we got them first as to how the dogs would take to them. The day we brought them home Ozzie was practically wetting himself with delight. I couldn't decide whether it was delight as in 'Hey, new friends!' or delight as in 'Hey, DINNER!' The first day I let them out of the coop, they strutted confidently around him and all was fine until one of them flapped her wings beside him. That was too much for him and he grabbed her in his mouth. I thought we had our first casualty, but he dropped her again just as quickly. Since then he's been fine with them – mostly. Someone once told me that a springer's instinct is to retrieve fowl during a hunt and carry them back to his owner without harming them, but I'm not so convinced of Ozzie's restraint. There was a curious incident one day when I went out to discover all four hens sitting on top of the hen-house in obvious distress, a load of feathers on the grass and Ozzie sitting there looking guilty and trying to pretend that nothing had happened. I will never know what happened but suspect he was involved somehow. We also have a young Labrador pup called Sam and he is actually quite deft when it comes to rounding them up. It's very difficult to catch a hen unless they are very domesticated, but dogs are far more agile than we are and when Sam chases one they seem to just give up and stand deathly still. He then puts his chin down on top of it and sits there until I come over and pick it up. He has never been trained to do that, incidentally (he has never been trained to do much of anything).

In the winter when the days are shorter and the weather turns inclement, things get a bit miserable for the hens. They still lay in the winter, but usually not as many – some people say you should put a light in their house and that will keep them all laying, but somehow that doesn't feel quite right. If they stop naturally in winter then it's probably best to leave them take the rest. And, besides, I wouldn't even know where to begin to get electricity to their house without electrocuting myself and them. The perch ladder was the apex of my engineering skills.

Most people agree that domestic hens will keep laying for well over five years but output starts to decrease gradually after about fifteen months. After approximately seventy weeks of intensive laying a battery reared hen's egg laying capacity will plummet and as she is no longer financially viable she will be slaughtered. She is not, however, finished making money for the producer just yet – the better cuts of meat (legs and breasts), while too tough to be sold as 'chicken', typically wind up in soups, pastes and pet-food. The carcass and whatever meat is left on it, is then minced to what is called Mechanically Recovered Meat or MRM – just about the nastiest, lowest grade food product imaginable and winging its merry way to a hot dog near you. We have a few old hens at this stage and while it's hard to tell who is laying and who is not, we suspect that they are not contributing much. I should probably make it my business to find out and get rid of the ones that aren't laying, but for the moment life's pretty sweet for post-lay hens down at Kelly's Fancy Fowl Retirement Home.

It's hard to love hens individually. They don't have person-
alities, really, but collectively they have given us lots of happiness. They run stupidly after us as we walk around the garden
and in the summer, when I'm cutting the grass, they follow in
my wake like seagulls following a trawler. It makes me feel like
St Francis. Our relationship with them is pretty formal – they
produce the eggs, we feed them. My niece, Katie, named one
of our first batch Dora and we could identify her because she
had darker feathers on her head than the others had, but that
was about as personal as we got. Every time my sister serves
up an egg to Katie, she will ask, 'Is this one of Dora's eggs?'
My sister says yes, even if it's not.

We had been keeping hens for two years before we got a
visit from a predator. Whenever you talk to people who keep
hens they always have a *fox story*. It seems that no matter what
fortifications you put up around your hens, the fox will get in
eventually. Barbed wire? Pah! Shards of broken glass? An
intruder detection system? The wily old fox can defeat them
all. Whenever I heard these barbaric tales I would reply, 'No
fox has ever troubled us.' Like the boy who cried 'Wolf'. The
secret of my success, if anyone asked (they never did), was
that we were always careful to lock them up at night. And
while dogs and hens don't particularly love each other, I think
the dogs did provide additional protection from the fox.

Then one week we were away on holidays and my mother
and sister were on hen duty. The day we came home I called
my mother from the airport to say we were on our way and she
replied, 'I have something to tell you.' Somebody has died, I

thought, something's happened. I felt sick. Now, not to trivial-
ise the terrible suffering of our poor hens, I was very relieved
to discover that the 'something' she had to tell me was that
three of our hens were gone and the fourth one, poor Dora,
was in bad old shape. A neighbour of ours had been walking
his greyhound in the fields near the house and it got into our
garden. All that was left of the three hens were little mounds
of feathers on the grass. Because we were away, the dogs were
also away in kennels for the week. I like to think had they been
there they would have got all protective and chased the greyhound away. But, on the other hand, they might have got all
frenzied instead and joined in the massacre. It's hard to turn
down a nice chicken supper.

My mother was heartbroken that it happened on her
watch. The fright nearly killed poor old Dora, who was the
sole survivor, and she had a broken leg. She wouldn't come
out of the coop at all for about three days and even then could
only hop about on one leg. My father-in-law told me I should
use the stick from an ice-lollipop to make a little splint for her
leg – I couldn't decide whether he was having me on or not, so
I decided against it. Each day I tried to steel myself to wring
her neck to put her out of her misery. It sounds easy, but how
could you do it to the poor old dear? She'd been through
enough. A very kind neighbour heard my tale of woe and
came around with two hens as substitutes. It's scary, but that's
what they are – substitutes. They really did look identical to
our own lost flock. That's a sort of a flaw that hens have – they
are so interchangeable. As payback for the yellow tennis ball

trick, I told Mrs Kelly that day when she came home from work that two of the hens weren't dead at all and had just been hiding in the ditches and had now returned! Halleluiah! (God, I'm cruel.) She didn't know whether to believe me or not and I had to 'fess up eventually.

The two new hens pecked at poor Dora unmercifully – so on top of her near-death experience and the prospect of me wringing her neck hanging over her ominously, she also had to deal with her two new companions not liking her very much. I felt dreadfully sorry for her. Initially I felt a sort of resentment towards the newcomers because of their shoddy treatment of Dora, but they immediately laid an egg each every day so the resentment was short-lived. And that's how it should be – it's all about the eggs after all.

Chapter 6

GIVING UP

I have found since I gave up sales and started writing for a living that weekends don't really mean anything to me. That might seem like a pity when you think about it first, but it's actually because I love my weekdays so much more than I used to. Weekends are still great – they are just not as great, relatively speaking, as they were back when my weekdays were shit. I used to have a sick feeling in the pit of my stomach on a Sunday afternoon at the thought of going back to work on Monday morning. In fact, if the truth be known, I didn't really like Sundays at all because there was always a faint but distinct smell of Monday off them. I was never, EVER, able to sleep on a Sunday night. I tossed and turned, fretted and stressed, thinking about the week ahead. That can't be right.

When I started writing first I decided that I should stick to my old routine in terms of the amount of time spent with my feet under a desk. I would get up with Mrs Kelly at 7.00am and make her breakfast. Then I would be at my desk by nine and do a regular workday, breaking for lunch for an hour and finishing at five or six. That lasted about a week. First of all, I didn't have enough work on, so I was spending a lot of time day-dreaming and increasing my circle of friends on Facebook. Secondly, I discovered that I was the only journalist in the whole known world working those hours. Because newspapers are printed late at night, the journalistic working day tends to start and end later than a regular working day (much like a music producer's). I also figured that since one of the key reasons I got out of corporate life was that I hated the routine, it was pretty silly to be dragging that part of my old life into my new one. I still get up at 7.00am with Mrs Kelly (because, honestly, if I start staying in bed in the morning it's a short and slippery slope to complete and utter ruin – staying in bed all morning, having a G&T at 3.00pm while watching Oprah, wearing saggy-arsed tracksuits around the house because they're just *so* comfortable). But, as my own boss, I allow myself the luxury of some flexibility in terms of working hours. I find this works quite well when you're self-employed because there's a control mechanism of sorts there to stop you taking too many liberties – ie you're only getting paid when you write. You never miss a deadline, of course; that would be just silly. You do the work when you have to. But that's the only rule that you need to stick to – apart from that, you can have a bit of fun messing with standard working hours. As I said earlier, you can go for a walk with the dog at 9.00am if the mood takes you. Take the morning off. Hell, take the day off. Of course, that means you might have to work late instead, or work on a Saturday, or whatever it takes. But it feels pretty good to be able to claim dominion over the seconds, minutes and hours that make up your day.

All of our lives have become hectic, but we are all guilty, to some extent, of believing that there is nothing we can do about it. This is simply not true. Time and time again, a particularly reprehensible old chestnut gets trotted out to explain why modern life in the western world is the way it is: we are cash rich and time poor. Why do we eat takeaways instead of cooking? Cash rich, time poor. Why do we have someone in to do the cleaning and the cooking? Cash rich, time poor. Why do we use a dog-grooming service? Cash rich, time poor. Why do we pay someone else to queue at the passport office? Cash rich, time bla bla bla. Why won't you stop and talk to me? *Sorry, can't do it – I'm cash rich, time poor.* It's a lazy old excuse and when you stop to think about it, it's a load of nonsense. People who are 'cash poor' really do have less cash than people who are cash rich. But 'time poor' people have exactly the same number of hours available to them each day as everyone else. It's what they *choose* to do with those twenty-four hours that makes them time poor.

The problem, of course, is that there is relentless pressure on all of us these days to cram our diaries full with activities. In my old job it was a cardinal sin not to have your life packed solid with an endless blur of travel, phone calls and meetings, one after another to infinity. And most people in business are the same. Stopping to smell the roses – well that's what we pay artists for. Business people, not so much. But it's not just in the world of work that this pressure comes to bear. Parents are run ragged ferrying kids around to an endless array of extra-curricular activities:

dance classes, piano lessons, football, ballet, basketball practice. They even have a word for this disease now: hyper-parenting, and documented symptoms: kids not having time to just be kids. If you decided to jump off the merry-go-round by shoving them out into the garden to play instead, you'd be accused of being a lazy parent, of not giving them opportunities, of not wanting them to succeed. Sitting around doing nothing really isn't acceptable these days (unless it's packaged and paid-for relaxation in the form of expensive treatments, spas, yoga) and we are all the poorer for it.

There was some structure enforced on my working week by the deadlines that I had with *The Gloss, The Irish Times* and the agency work. I was having great *craic* with the restaurant review for *The Gloss,* which I was writing under a pseudonym at the editor's request. I messed around with the names of some well-known food reviewers, to see if anagrams of their names would suggest an interesting *nom de plume*. Eventually, stifling the chuckles, I came up with Lord Toomey, which is an anagram of Tom Doorley, über-reviewer, *Irish Times* contributor and all-round top bloke. Given that Tom is an expert on food and I'm clearly not, this seemed to fit nicely, though I'm pretty sure the vast majority of *Gloss* readers never noticed my clever self-send-up, which in some ways defeated the purpose of the whole thing. Anyway, the name Lord Toomey immediately suggested a cast of characters to me, so each month I staged a little mini-play involving him and an array of fabulous dinner companions. It was a lot of fun (for me at any rate). Basically, I cast him as a lecherous,

Above: Dunmore East, County Waterford.

Below: A view of the 'leaky little cottage' from the end of the garden. The same view that our pet bat had following his eviction from the house.

Above: The seedy side of town – the polytunnel in early summer.

Left: 'Hmmm, I wonder what these are?'

Opposite top: DIY gone mad… Roger with his harem in the fore-ground, various DIY efforts in the background.

Opposite bottom: Obviously I was still working in sales at this time! But the hens didn't mind.

Above: Roger stalks two unsuspecting bantam hens – he uses stealth and cunning to get his wicked way.

Left: A hen stands at the henhouse door pondering life's great mysteries, like: 'How do I get down from here?'

Opposite top: Just a few days old, a fluffy, yellow chick is already out and about exploring the run. Could you really eat something this cute?

Opposite bottom: Libby checks out the pigs – and they check her out. A meeting of minds?

Above: Life is beautiful —
piggies soak up some rays.

Left: A dog's life is
pretty good too —
Ozzie and Sam.

Above: The Tamworth Two in their pen shortly after they arrived – note the des res.

Right: The fancy mower with the redneck stickers – weapon of grass destruction.

Left: Charlotte and Wilbur strut their stuff for the camera.

Below: Mrs Kelly with Ozzie.

permanently sozzled old English toff, who used to work in the diplomatic corps and liked to boast about the fact that Winston Churchill was his grandfather's cousin. I never fully explained why he had retired to Ireland and how he had landed a plum job reviewing restaurants for *The Gloss*. Nobody in Lord Toomey's universe seemed to work and they always seemed to be drunk, stoned and permanently disaffected. His dinner companions included his liquored-up friend Dolly, his illegitimate daughter Violet and her coked-up friends, his equally lecherous and sodden cousin Lord Hinchingbrooke, an old colleague from his Foreign Office days called Old Major, and an ex-lover called Clarissa Marling, to name a few. I enjoyed my monthly trips to Lord Toomey's world which was elaborately constructed to disguise the fact that I knew feck all about food.

If writing the restaurant review was fun, the work I was getting from the freelance agency increasingly was not. The idea behind these agencies is essentially that they send you an idea for a feature, which you then write, and they pitch it to various newspapers. The catch is that the copyright to the article passes over to them once the piece is written, which means they can do pretty much anything they want with it. Sell it to five papers. Or ten. Chop it in half. Sell some of it. Sell all of it. Re-write it (even though your name is still at the top of it). When I started with them first, I didn't really care where the articles ended up, but a couple of bad experiences gave me a useful lesson in how important it is to be published in publications that you respect.

The first commission I did for them was a thousand-word feature on song-writing inspiration. I picked five iconic songs (eg The Beatles' 'Strawberry Fields', Carly Simon's 'You're so Vain') and included an Irish angle by interviewing Glen Hansard from The Frames about 'Revelate'. The editor assured me that it was going to be published in a national paper in one of the weekend supplements, but a few weeks came and went and nothing showed up. Finally, to my absolute horror, a short 150-word paragraph from the overall piece (the bit that focused on the inspiration for 'Revelate') was published, buried deep within the rancid bowels of one of the trashier Sunday tabloids. Incidentally, I got paid as soon as I submitted the piece, so it shouldn't have mattered to me where it got published or, in fact, whether it got published at all. But, it did matter to me. I was pissed off. I was pissed off that the piece was butchered the way it was and, ok, I'll admit it, I'm a newspaper snob. I don't much like tabloids and I really don't like Sunday tabloids.

Anyway, I guess when I started with the agency I hadn't even considered the possibility of the pieces being published in a tabloid. I suppose I figured that since I write in a broadsheet style (wordy, morose, hefty, weighty, ponderous, someone wake me up when it's all over) as opposed to a tabloid style (SNAPPY! CRAPPY! SLEAZY! LOTS OF CAPITAL LETTERS, *ITALICS* AND PUNCTUATION MARKS!?!) that the agency would default to broadsheets when deciding where to send my features. Clearly, this was very naive on my part. I never considered for one second that they would go to the trouble of actually rewriting a piece in tabloid style and then submit it to a tabloid with my name on it. Well, wouldn't you know, that's exactly what was happening.

Another feature that got the tabloid treatment was about the Irish launch of the Multiple Sclerosis drug, *Tysabri*. It was a serious piece and I included interviews with a director of Elan (the company that produces the drug), representatives from MS Ireland and a young MS sufferer called. She was a really nice girl and we had two long chats about her experiences. It was actually the first time she had talked to any journalist and she was worried that the piece would sensationalise the dangers of being on *Tysabri* as opposed to emphasising the benefits. That was the first time that I came face-to-face with people's negative perceptions of journalists – that we basically make up shit and write whatever we think will make a story sell. I promised her I wouldn't do that.

The editor at the agency came back to me and said that while it was a nice piece, he had to rewrite it in 'tabloid style' because one of the English daily tabloids was interested in it. I felt queasy. The feature started with the immortal lines (cue scary mood music): 'It started in a laboratory in southern San Francisco in the early nineties. Paralysed laboratory mice were suddenly able to walk when given a drug. Around the same time, thousands of miles away in Dublin, a fourteen-year-old girl noticed something wrong with her sight. Only a controversial treatment offered hope ...' Oh sweet Jesus! I mean, it sounded like the script for some dodgy made-for-TV B-movie. I felt terrible. I was thinking maybe I should call her

– explain that the piece I wrote was a factual, fair, un-sensational piece of journalism rather than the piece of shite that got published. But I didn't call her. It wouldn't have mattered. My name was against it, so I was responsible.

The straw that broke the camel's back was a feature about a young Irish entrepreneur who set up a new TV channel, Bubble Hits. His story was a good one because (a) he was only twenty-three and (b) he was making broadcasting history because it was the first music channel without ads. I wrote a thousand-word piece on him, having interviewed him twice. I thought he was a nice enough guy, a little cocky, perhaps, but then again if I had been a millionaire at his age (like I'd planned) I would have been a complete tosser. Again, the piece showed up in a tabloid under the headline 'The Boy Who's Forever Blowing Bubbles'. This time, someone (either the agency or the newspaper) had completely rewritten the feature – even though my name was still over it as the writer. I couldn't help thinking – what's the point of all this? Why do they bother paying someone to write something and then go and rewrite it themselves anyway? It just seemed like such a waste of time. But that wasn't the worst part. While my piece had been a fairly benign, respectable treatment of the guy (I had nothing but admiration for him, really – I mean, when I was his age I was basically drunk all the time), this was a complete assassination. This Michael Kelly didn't like the young man at all, calling him 'an irritatingly cocky upstart who will no doubt be successful but will never stop being fundamentally tiresome.' Huh? What if I meet him in the future, what will I say to him? What if he decided to sue for libel? Who would he sue? Me? The paper? The agency? I wasn't sure and that fact alone really, really bothered me. Basically I thought: Sod this. I've come this far, I've gone through all the trauma of jacking in my job. I didn't do it to have my name put against any old piece of shite. If you don't have your good name, you have nothing.

As it happened, with the agency commissions out of the picture, I had more time to spend trying to make progress with the *Irish Times*. The editor in the magazine asked me to take over a column called 'My Big Week'. I was delighted to take it on and at the time of writing, I still do it. Basically, it's a short piece on someone who is having a big week. It can be tough to come up with the person each week and to keep it varied for readers, and the deadline does appear to come around relentlessly, but, on the other hand, the fact that you get to talk to all manner of interesting people more than makes up for it. I love it. Sometimes the person having a big week is famous, more times they are just ordinary people involved in extraordinary events. Since I gave up my job a few people have said to me: 'Don't you get bored sitting in your office writing away all on your own?' How could I get bored when I have such an interesting window on the world?

Shortly after that I started writing the 'What's On' segment in the magazine, which is a sort of round-up of the week ahead, covering arts, music, sport, events, festivals. It appears on the same page as 'My Big Week', so I sort of feel like I have that page to myself each week, which is nice.

My e-mail address gets published at the bottom of the article so pretty much immediately after I began writing it I started getting bombarded with e-mails from basically anyone in the country running an event. I didn't have broadband at the time and my dial-up internet connection literally groaned under the pressure. I like frenetic e-mail activity – it's a great way to keep the old finger on the pulse of what's going on in the country and it also throws up lots of ideas for other features. It also makes me *feel* busy. A fellow *Irish Times* contributor told me once that our job is to act as a conduit to bring interesting stuff to the masses so to speak, and that's a nice way to look at it. The challenge is to discern what actually constitutes interesting stuff.

I made some headway too with the editor of the *Irish Times* Tuesday health supplement and a feature there called 'A New Life', which is essentially interviewing people who have made a major change in their lives and careers, exactly as Mrs Kelly and I did. Talking to people like that and then condensing their life-story into a thousand words is phenomenally interesting work.

One of my proudest moments since I started freelancing was getting my first 'Irishman's Diary' published on the letters and opinion page in *The Irish Times*. 'The 'Irishman's Diary' has been a feature of the paper since 1927 and is, to my mind, the prime retail space in the entire newspaper. The column was synonymous with Kevin Myers before what might be called his 'little bit of bother' and his transfer to *The Irish Independent*. Frank McNally now writes the piece Tuesdays to Fridays, which just leaves Mondays and Saturdays up for grabs. I've had, maybe, half-a-dozen of them published, mostly on environmental issues. This doesn't exactly make me a regular, but I have to admit that on the days when I've opened up the paper and seen my name under the famous heading I've felt an unmistakeable thrill each time. These are the things that keep you going. My mother was Kevin Myers's number one fan and bitterly lamented his defection to the *Indo*, so I think it gives her a good buzz too.

In 2007 I started a column for *The Irish Times* magazine called 'Giving Up'. It was about, not surprisingly, giving up things. The genesis of the column was that when I gave up my job, I also gave up lots of other things which had been part of my life – ironing shirts, wearing ties, daily shaving ... So I pitched the idea to the editor for a weekly column where I would give something up and write about what it was like.

There was quite an unexpected side-effect to the whole process – I actually found that I enjoyed giving things up. It's similar, I guess, to the perverse pleasure you used to get from good, old-fashioned Catholic self-sacrifice and abstinence, particularly at Lent. There was always something to be learned from giving something up, even if it was temporary. And the timing of the column was perfect for me – having just given up my job, I was in the right frame of mind to be a bit, well ... bohemian. The feedback I got from people about the experiments was interesting too – the idea that you would deliberately deprive yourself of something you enjoy is considered to be sort of 'bad form' in Ireland these days. Almost subversive.

THE CAR

My first experiment was to give up my car for a week. In my
last year in sales I put up about forty thousand miles travelling
around to customer meetings, spewing out over ten tonnes of
greenhouse gas (that's the equivalent of eight hundred
wheelie bins). Like so many other people in Ireland today, my
car had become a sort of surrogate home – a place where,
very often, I would spend entire days. I also spent God knows
how much money at service stations buying petrol, getting my
car washed and buying shite forecourt stodge – hash browns,
chicken wings, potato wedges, breakfast rolls, crappy sandwiches, endless cups of tea and coffee, packets of crisps, bars
of chocolate. The bright neon of the service station is the
perfect lure for a new generation of forecourt commuters –
sales people, truck drivers, weary commuters – all bored of
driving, and looking for grease. We're like vampires; we step
from our cars with gaunt faces and bags under our eyes, fill the
forecourt bin with the detritus of weeks of automobile dining
– waste packaging, styrofoam cups, the core of a rotten apple
or a few banana skins. We approach the counter like zombies to
pay for our goods, and then return with our meal to the comfort of our cars, our brief liaison with the outside world over. A
lot of the time when I got back into the car, I would turn on the
engine and drive away immediately, balancing a sandwich on
my knee and taking occasional slurps of tea along the way –
admittedly, a fairly cavalier attitude to road safety.

In that context it seemed only right and proper to treat
myself to an entire week free from the car. The rules were that

I wasn't allowed to drive or be in a car (getting a lift to the shop was cheating). We live about three kilometres from Dunmore East, so each morning I cycled to the village for the paper. I hadn't cycled in years so it was part nostalgic reunion and part arduous struggle. OK, mainly arduous struggle. But it was enjoyable to rediscover the art of travelling somewhere and trying to actually relish the journey itself. We forget the fact that there was a time in the not too distant past when people actually enjoyed travelling – as in the journey itself, as opposed to the destination. Mandatory commuting has robbed us of this particular pleasure. Most of us hate journeys now because they are what separate us from our lives. But it was not always thus. CS Lewis talked about the car 'deflowering the very idea of distance', which I think is a wonderful way to put it. It's like that guy I heard on the radio who is commuting from Kilkenny to Dublin – it's a 120km journey, but the car makes a mockery of it and turns it into something commutable. Giving up the car for a week confined me to a relatively small realm (basically only as far as I could carry myself on foot or on a bike), but, ironically, that realm seemed roomier and far more interesting as a result. I walked out our gate one of the days and I realised it was the first time I had ever *walked* out my gate. I had driven and more lately cycled. But never walked. Imagine that?

Out for a cycle one day, about half a kilometre from our house, I discovered a huge memorial stone at the side of the road, dedicated to a soldier who was killed in a road accident there in 1948. I've passed that a thousand times in the car and

I couldn't believe I'd never noticed it before. A little further along the road, two horses in a field beside the road trotted over to say hello, and I stopped for a while and we had a chat. They seemed happy with the attention and a rub on the nose and I was happy to stop (anything for a break from the searing pain in my calf muscles). 'The modern boy,' said Lewis, 'travels a hundred miles with less sense of liberation than his grandfather got from travelling ten.' What would Lewis have made, I wonder, of kids sitting in the back of cars watching DVDs on TV screens that are built into the backs of the front seats? Ambling along on my bike, with cars flashing by me in a terrible hurry, I was taking in things I never take in while driving, and thinking about how much of a rush we are always in.

Anyone who knows Dunmore will be familiar with the fairly serious hill from Killea church down to the village. The first day I cycled to the village I descended that hill at what felt like 250kph, feeling a mixture of adrenalin-induced euphoria and absolute terror – at one point I had to brake hard when I caught up with a pair of nuns in a Micra. I was smiling so hard when I got to the bottom I looked like I had a chronic vitamin C surplus, my eyes were streaming at an alarming rate and there was snot pouring uncontrollably from my nostrils. It felt exciting and also kind of familiar – I later realised it was that intense adrenalin rush, that feeling of elation, of being alive that you had frequently when you were a kid, from cycling down a hill with no hands or just after you'd rung a doorbell and were in the process of running away. For every freewheel thrill-ride downhill, there is, of course, the corresponding

'King of the Mountains' slog back up. I was in a heap afterwards. It does make you wonder why we pay through the nose for gym membership, driving miles to get there in the first place and exercising in an unhealthy, air-conditioned building when you can pan yourself out on the bike for free and get some fresh air into the bargain. Giving up driving means combining travel and exercise into one activity.

Another day I took the bus to Waterford and I was left wondering why I had never thought of it before. For a start, there is a bus stop about a hundred metres from our house, then it costs only €2 to get to Waterford. *Two euro!* Nothing costs €2 anymore! The bus drops you right into the centre of the city, so there are none of the parking hassles you normally associate with a trip to town. There were a few, mostly old, people on the bus chatting amiably about life in general and I enjoyed the feeling of being part of their leisurely world for a little while. While I was waiting for the bus, a delightful old lady stopped and offered me a lift, but since that would break the 'Giving Up' rules I had to reluctantly decline. This is what other people are up to when we're all busy killing ourselves working – lolling around on buses, shooting the shit, giving lifts to strangers.

THE MOBILE PHONE

My next 'Giving Up' experiment was to power down my mobile phone for a week. This was interesting, given that in my old job I had suffered from a fairly serious BlackBerry addiction (the handheld e-mail device, not the fruit). I argued vociferously for a BlackBerry, so much so that the managers gave me one in the end just to shut me up; and, as if by way of having their revenge, I gradually became enslaved by the thing. The network providers who sell them promote them as a device that will set you free and as a way for stressed-out executives to regain control of their lives, to move the needle on the old work-life balance scales. It's complete nonsense. If anything, it pushes the needle back the other way – back towards work, back towards your employer. I actually took the BlackBerry on holidays with me one year. Imagine that? I would power it up at night to have a sneaky look at the e-mails, reasoning that it would lessen the impact of that 'first Monday back' e-mail onslaught. Jesus, I'm almost embarrassed to write that now. Handing back my BlackBerry when I resigned was a symbolic moment; it felt like the moment I stopped being a corporate slave. It was also the moment that a relative peace descended on my life, which was rather eerie at first.

In my old job my phone used to ring and beep incessantly, almost 24/7. On busy days I could spend literally the whole day on the phone and the side of my head would be hopping by the evening time. These days my mobile rings a handful of times on a busy day, and some days it doesn't ring at all. For the first few months I was still in BlackBerry mode, checking my new mobile every couple of minutes for messages. There were none. I felt a curious mixture of delight and sheer terror.

Leaving the office meant having to buy my own phone, and because I wanted to keep my outgoings down I decided to get a pre-paid one, rather than have a huge monthly bill. They really keep your bills down because you are constantly aware of your credit running out and you tend to be more careful about your mobile use. On the other hand, they tend to make you feel like a teenager rather than a fully paid-up member of the adult world. A friend of mine slags me that I am the oldest pre-paid phone owner in Europe by about fifteen years. The first time I got a message on my new phone saying 'You're running low on credit' I felt all cheap and nasty.

Mobiles are tough taskmasters. For a small, inanimate object, they have a disproportionate level of control over us. When the mobile bleeps or beeps or rings, it expects an immediate reaction. And it usually gets it. Say, for example, you are upstairs in your house doing something and you hear your phone beep downstairs to signal the arrival of a text message. How long will you leave it before you go down to check it? A few minutes? Half an hour? When you're god-damned ready? Or do you drop everything and sprint down the stairs immediately? And when you finally pick the phone up and check the message is the text ever as important as the immediate attention you gave it? Never. *How r u. Where r U. Any craic?* The greatest lie that has been peddled to us is that the mobile has turned us into great communicators. In fact, paradoxically, it's an isolating device. When you're on a mobile you're in a private world, ignoring the real events going on in the world around you. The mobile plays to our innate shyness – we can get in touch with people without having to meet or even talk to them. *I am gr8.*

The week I gave up my phone I was surprised to find that I felt very disconnected and very uneasy, like I was missing a limb. I counted all the times I reached for the phone to make a call or send a text, and it was scary how often I did it. What if something happens and someone needs to contact me, I worried. I had to soothe myself with rational thoughts: don't be so silly, in all likelihood, nothing is going to happen. And if it does? Well, if it's that important they'll find an alternative method of contact, just like they did before mobiles were invented. The other thing that you discover is how frequently we use texts as a lazy method of contacting people. Take this example: my nephew was starting school for the first time that week so I went to send a text to my sister to wish them luck. But my wife was having none of it. 'That's cheating,' said Commandant Kelly. So I picked up the landline and called my sister. A text saying 'hope things go well 2moro' would have taken twenty seconds to send. The phone call took twenty minutes. But I was, of course, the better for it. By the end of the week I was obsessed with how many texts and messages were stored up on my phone which was sitting forlornly in a drawer in the kitchen. In my old job, a week away from work usually meant there would be two hundred e-mails and fifty phone messages waiting for me when I got back. In my brave new world, when I powered back up, there were seven texts. Four of them were from my mother. Bless her.

THE TIE

My other great bugbear from corporate life was having to wear formal attire every day – suit, shirt, tie, formal shoes. I really can't explain the childish glee I get from being able to slum it in jeans and trainers. Years ago, I calculated that based on an average iron-time of ten minutes per shirt and wearing a shirt a day for forty years I would spend seventy-three *days* of my life ironing. What a phenomenal waste of a precious life. Imagine what you could achieve in seventy-three days? Imagine that I bothered to sit down and work that out? I like dressing up sometimes – it's nice to cut a dash – but when you have to do it every day, it's different. It's repression. It's also very false in some ways. Back in corporate land, I was always trying to make an impression with my clothes. The day I went to my first interview for the sales job at the age of twenty-two (and looking all of fifteen) dressed in a brand new, navy three-piece suit, I was trying to impress. Throughout my decade in corporate life I used smart business attire to *appear* successful, dynamic and business-like. But now I just wear what I wear and I like to think that this means I am altogether happier in my skin and not so concerned with how I appear to other people. But maybe I am reading too much into it.

Revelling in my new-found freedom, I gave up wearing neckties to formal functions and wrote about it for the column. Neckties always represented the corporate noose around my neck – can there be any greater abomination on God's earth than the necktie? Having to strangle yourself and virtually cut off the blood supply to your head in the interests of … well, what exactly? Ties have no function. They are simply a badge of conformity, of belonging. Of membership. They are the ribbon on a gift – completely useless but thought to complete the wrapping. I swore when I gave up my job that I would never wear a tie again, but was surprised to find that society still expects them and can get quite sniffy if we don't play ball.

I tested that premise at a wedding that Mrs Kelly and I attended. As I dressed myself that morning I was brave. Defiant! *Damn you to hell, tradition. Be gone, formality!* But as we got to the church my resolve waned. I felt like I was wearing a pair of shorts and a t-shirt. I wouldn't mind, but I had invested considerable effort in trying to ensure that the rest of the ensemble would be sufficiently smart to take attention away from the fact that I wasn't wearing a tie. I was wearing a nice pinstriped suit and some decent shoes and a fetching pink shirt. Nothing says high fashion like pink. It is thought by some psychologists that the wearing of a tie is a subconscious effort by men to draw females' eyes down towards the male genitalia. As I was tie-less, the evening was unfettered by bothersome women staring at my crotch. There were a few others at that wedding not wearing ties, so the experience wasn't too bad, but a few weeks later we had another wedding to attend and, emboldened by my first success, I decided to go naked again. This time I was the only guy at the wedding not wearing a tie. I felt like a beatnik. At our table, there was a couple who had clearly colour-coordinated their outfits. The woman was wearing a fetching black dress which had some orange flourishes in it. The guy had an orange tie on. Posh and Becks would, no doubt, be pleased, but for us males this is a scary development. Are we now a fashion accessory – occupying a position of importance somewhere between the hairclip and the ankle bracelet? I tried to explain this to a friendly girl sitting on my left. 'Well, at least he wore a tie,' she sniffed.

MEAT

I gave up meat for a week, which should have been easy, but wasn't, given that I come from a family of notorious meat-eaters. When I go over to my mother's for dinner and it's just the two of us, she will usually cook two chickens ('Just in case someone else joins us') and we enjoy a gluttonous feast of ancient Roman proportions. Part of the joy of cooking with meat is that it's so simple. You just pop your head in the freezer and pull out a few lamb chops, or perhaps grab some mince and rustle up a few homemade burgers. Simple. But at the start of my planned week without meat, I sat down to think about what we could eat, and couldn't come up with a single vegetarian recipe. Does soup count as dinner?

Having trawled through various cookbooks I came up with a vegetarian meal plan and glumly went shopping. 'Hi there, Michael' said my friendly local butcher, smiling and reaching for his favourite sharp knife. I walked right past the meat counter with my veggie-laden trolley, trying to avoid eye contact and feeling adulterous. By the end of the week without meat I felt pale, lethargic and barely able to stand up – ah no, only messing. It was fine being off meat for a week, but I don't think vegetarianism is for me. I'm a Kelly, after all. Incidentally, after that article was published I received my first hate mail – well it wasn't really hate mail, more a mild-dislike mail. It was from a vegetarian and she took issue with the fact that I included fish on the menu when, clearly, fish isn't a vegetarian option. You can't please all the people, all the time.

DEODORANT AND SHAVING

After that, I had some fun with personal hygiene, giving up deodorant and shaving. I got lots of *interesting* feedback on the deodorant one. I suppose people thought it must be pretty disgusting. In my defence, I did find an alternative of sorts, so I wasn't a complete minger for the week or anything. And I did have a real issue with these products. Pretty much every morning I wonder about the health implications of using deodorants. They have been the subject of various, largely unproven, health scares over the years; it was suggested, for example, that the aluminium in deodorants could cause cancer and Alzheimer's. I'm no scientist, but, bottom line, could it possibly be good for you to spray a concoction of chemicals onto highly porous skin? Take a look at the ingredients on your brand – mine showed up the following, among other things: butane, propane, aluminium chlorohydrate. Propane? Isn't that what they use in patio heaters? Why am I spraying that on my pits? What if someone lights a match near me?

For the first few days *sans* deodorant I was disappointed to discover that I didn't seem to sweat or smell at all. I am not sure if that is a comment on how unnecessary deodorant is or on my sedentary lifestyle. On the third day, following some activity in the garden, I was excited to find that I was actually sweating – I wasn't falling over with the stink or anything, but it was a trifle unpleasant all the same. So I found a recipe for homemade deodorant on the web. This is what I'm reduced to, I thought as I cooked up a brew of oils (almond, rosemary, lemon) and beeswax. The finished product was a jar of pungent, soft wax, a very tiny amount of which is applied to the underarm. I was pretty excited about my homemade deodorant to begin with, and could even envisage making up large batches to give to people as quirky Christmas presents. But it didn't take me long to realise that homemade deodorant has a slight flaw: it doesn't work. It tends to let you down just when you need it most. Most articles that I read on giving up deodorant seemed to suggest that it takes time for your body to self-regulate – if you are willing to endure a few months of stinkiness then eventually you will stop smelling. Or perhaps you just stop noticing? Anyway, I wasn't willing to endure the few months and went back to my own brand almost immediately the week was up.

Mrs Kelly raised her eyes to heaven when I told her that my next plan was to embrace hippiedom and give up shaving. I hate it when big corporations take over (or invent) a daily routine and then make you pay through the nose for it, like some form of tax. Shaving is a good example of this. ('One-a-day' advertising for pro-biotic drinks is another. Do they really think we can't see through that ruse?) Razor-blade manufacturers have completely hijacked the process of shaving – they are like a toll-road operator sitting in a little booth on a road that should be free travel, sticking a fat paw out the window to take money off us every time we pass by. They do this by regularly changing the razor design and then phasing out the accompanying blades to force you to upgrade. Did you know that an eight-pack of blades costs €16? *Sixteen euro!* For feck's sake. That's extortion. Increasing competition between the leading players in the industry has made the whole razor thing ridiculous. For one thing, there's that bizarre drive to see who can fit the most blades on the end of one razor. It used to be that two blades was the pinnacle of razor technology, then it was three. Now it's four, with a 'moisture strip' – what the hell is that? Where does it all end? Edward Scissorhands calling round to your house to shave you? And then there are those overtly macho ads (guy drives motorbike with semi-naked blond model on back) and the even more macho product names. Mach 3! Mach-3 Turbo! Mach-3 Turbo Power Fusion Force 5! (OK, I made the last one up, but you know what I mean). Shaving is macho. We get it!

I didn't particularly like having a hairy face but I did enjoy the freedom from the daily routine. A few days into it, the stubble was actually pretty cool in a sort of Ryan Giggs kind of vibe. But after day three or four it started getting really itchy and I couldn't stop at it, massaging it like a beatnik pondering the mystery of life. Alarmingly, some of the growth was actually of a reddish hue. 'Beat it, Ginge,' said Mrs Kelly when I tried to kiss her. On the fourth day I actually got a spot below my lower lip and I felt like a pubescent teenager. The gradual growth means you don't get any great sense of change and it's only when you meet someone who has only ever seen you clean-shaven that you get a sense for just how awful you look. I met my sister towards the end of the week and she actually laughed in my face, which was quite off-putting. 'It puts ten years on you,' she said, and then laughed some more. I happened to catch sight of myself in a shop mirror and it scared the living shit out of me. I looked like one of the Dubliners. By the end of the experiment I couldn't wait to shave. You never feel really clean with a beard. I would get out of the shower and feel like I shouldn't have bothered. So, reluctantly, it was back to the Gale Force-5 Xtreme Powers for me and complete submission to the Gillette master.

THE DISHWASHER

One Sunday we had some family over for brunch, which I cooked on the barbecue. Afterwards I filled the dishwasher with greasy plates, knives and forks, mugs and glasses, the grill from the barbecue and various other bits and pieces. Then I realised that this was the day I was supposed to start giving up the dishwasher for a week, so I had to take them all out again and start scrubbing. Stupid bloody column. I was fairly annoyed to discover too that giving up the dishwasher was in fact *bad* for the environment – if used correctly, ie full, dishwashers, apparently, use less water and energy than you do when washing by hand (and to make matters worse, I was putting about five tea-towels a day in the washing machine). I guess I could have tried to improve things by using the dish-water for watering plants, but do people really do that sort of stuff? Never mind, it was a victory for the soul. Mrs Kelly and I chatted amiably as I washed and she dried after dinner each night. You wouldn't do that beside the dishwasher – well, you might, but it would be just weird.

*   *   *

I startled my neighbours by showing up on their doorsteps to introduce myself during 'Giving Up Ignoring the Neighbours' week. I gave up alcohol in the middle of the Christmas party season and found that making chit-chat while sober is actually quite a difficult skill. I tried and failed to be a better person when I gave up judging people for seven days, and I filled a swear jar the week I gave up swearing. I endured a self-imposed news blackout for a week (interesting one for a journalist) and found myself dreaming of 'Questions & Answers'. I gave up exercise one week and the internet another (just how *did* journalists do their research before the arrival of the World Wide Web?). I got profoundly irritable while off coffee and tea, and took up clove-sucking the week I gave up chewing gum. I gave up shaving-cream permanently after trying out baby-oil as an alternative to crazy expanding foam (I promise you it works, try it – you'll never go back).

In the interests of complete honesty, here are a few things that were on the list of things to give up but which I didn't ultimately try on the grounds that they were just too impractical (ie I didn't have the balls or Mrs Kelly wouldn't let me): giving up showering/bathing; giving up talking; giving up sex; giving up getting out of bed. I did manage to give up TV

– and figured that a week would be just too easy, so I waited until that great feast of self-sacrifice, Lent, and gave it up for six weeks.

People were almost offended by the idea that someone would deliberately give up TV. 'Why on earth would you give up telly?!' they asked, standing protectively in front of it. We are not huge TV people in our house, but I guess it does get switched on every night – mostly we just sit there flicking between channels wishing there was something good on and watching re-runs of *Friends* or *Scrubs* that we've already seen a hundred times. Given that I spend so much time complaining that there's nothing on, you'd think I would have been happy to give it up for forty days. But no. I was actually kind of apprehensive. It's that niggle at the back of your mind that you're depriving yourself, senselessly, of something nice. Something good. A little treat to help you chill after a hard day's work. (Oh, who am I kidding – writing's not hard work!)

In the first week, I went to meet two old friends for a pint to catch up (thank God I wasn't off pints too), thinking that going to the pub would be a good way to take my mind off the fact that I couldn't watch TV. I should have known better. There were two TVs on in the pub, one at either end, showing two different Champions' League matches. So I sat there over two or three pints feeling guilty and trying to avoid eye contact with my nemesis. For the record, Chelsea won. That's one thing I've noticed: TVs are everywhere. You're not aware of that until you're studiously trying to avoid them. Queue at a bank or a newsagent's and everyone in the queue is staring blankly at Sky News on a 182-inch flat screen plasma over the counter (a clever ploy so that you don't notice your lunch break being frittered away).

Mrs Kelly arrived home one day with *Cheaper by the Dozen* on DVD. We had a discussion on whether a DVD would break the TV ban. I reckoned it would. We had another discussion on whether *Cheaper by the Dozen* is such a crap movie it might count as penance. Most nights it didn't really bother me all that much that we couldn't switch on the telly, but on a few occasions I really felt I was missing out. For example, Ireland's crunch Six Nations' match against England had to be listened to on the radio – it wasn't bad, but it's not exactly a multi-media entertainment extravaganza. However, I did move the radio from the kitchen to the sitting room and took to listening to it in the evening. There's a whole world out there on night-time radio, which was a bit of a revelation. Sitting there in front of the fire, I experienced a whole 'listening to the wireless during World War II' kind of a vibe.

I also got into really reading the papers. I'm a daily newspaper reader, but tend to be a scanner – I read things that interest me. But with nothing else to do I started reading the entire paper, cover to cover. That was a hoot. I also got through lots of books. Mrs Kelly and I engaged in an activity called 'talking' a lot in the evenings. I was reading about this guy on the internet who gave up TV for good. Now, there's an achievement. He said that about six weeks after he stopped watching it, he finally realised what a waste of time TV really is. I'm pretty sure I'll never have an epiphany like that – once

Lent was over I was immediately back to my *Scrubs* re-runs and vowing that one of these days I'm going to throw the telly out the window. But I never do. I did have a realisation of sorts; TV's great achievement (or not so great, depending on your point of view) is getting you informed quickly, with a series of quick sound-bites while you lie on the couch in a sort of coma, letting it all wash over you. But to get yourself informed when you don't have TV? It's hard work. You have to read. You have to listen. And maybe, just maybe, you are better informed as a result.

Chapter 7

MEAT IS MURDER

Having established our bona fides with the hens, we thought about rearing chicks next. Apart from the general cuteness of having little fluffy chicks about the place, we thought it would be useful to replenish our stock of laying hens, then maybe keep a few for the table. Chicken, along with pork, is a meat we have all but stopped eating over the years due to concerns over the impact that intensive farming has had on these creatures' miserable lives and the subsequent impact on the quality of their meat.

The discerning meat buyer can pick up decent beef and lamb in the shops and be relatively satisfied that they are getting Irish produce that has spent much of its life in the open air, happily munching on grass. You can have no such confidence about pork or chicken. There's a pretty basic equation that you can apply when it comes to buying chicken: the cheaper it is, the shorter and more miserable the bird's life has been. It takes a hundred days or so to rear a really good chicken, which is why producers need to charge more for them, whereas ordinary chickens are produced in thirty-five to forty days, which is how suppliers can afford to sell them for €4. It's fairly obvious which one will taste nicer.

Our local supermarket stocks organic chickens reared by a guy called Paul Crotty here in Dunmore East. Organic is very different to free range even when it comes to the humble egg, as we know, but when it comes to the chicken meat itself, it's worlds apart. Remember that free-range regulations do not have anything at all to say about what the chicken is fed (they only concern themselves with how much space the birds have), and the feed is the most important thing of all when you're going to be eating the animal. If it's been fed crap, then you are eating crap. If you are buying free range, you need to check on the packaging what the birds have been fed. Organic chickens typically cost upwards of €15, which is expensive when compared with the €4 alternative sitting on the shelf beside it. But that doesn't make it dear. It just means the €4 is too cheap. Yes, people have families to feed on a limited budget, but buying a chicken for €4 is not the answer. In my opinion, even in my current circumstances when I earn far less than I used to, it's worth shelling out the extra money for a really good chicken and then trying to make the most of every last morsel of it. There's no point in giving out about the price of free-range chickens and then cutting off the breasts and firing the rest in the bin, as some people actually do!

I discovered that one of the most useful skills you can learn when it comes to being thrifty in the kitchen is how to joint a chicken – you can use the breasts for one meal, the legs and wings for another (a nice casserole, for example) and the carcass for a stock or soup. It takes less than five minutes to prepare the best stock imaginable from the carcass – fire it into a pot with some onion, celery, carrots, herbs, water and seasoning and boil it for three or four hours and you're done. The stock can be frozen, and will make you a fantastic soup, or can be used for the casserole. Sometimes, if I'm feeling particularly thrifty (or mean, depending on your point of view), and as a final bid for not wasting the precious bird, I take every last morsel of meat from the cooked stock bones and make a chicken curry with it. Otherwise I give the scraps to the dogs and they love me for it.

My mother-in-law keeps about twenty or so chickens at a time and then kills them all at once so she has a big batch of them in the freezer. She buys them as chicks, which saves all the bother of actually getting the eggs to hatch. We thought about doing the same, but then, around Easter-time, we got all seasonal and sentimental and reckoned it would be nice to have little chicks about the place. The key to producing your own chicks is to get a cockerel or rooster to fertilise the eggs. In some ways, it's no harm to have a rooster in your flock of hens, anyway, even if you're not interested in rearing chickens for the table, because they tend to chaperone the hens around the place, keeping them all together and (hopefully) keeping them out of harm's way. Cockerels are quite chivalrous too – if he finds a worm or some other morsel of food while he's foraging, he will let out a distinctive *Bok bok bok,* which brings all the hens running over to him and he then stands back while they eat the find on him. Roosters are great fun to have around and open up the possibility of all kinds of sexual innuendo: Where's your cock? That's a fine cock you have. Can I hold your cock? How many cocks do you have? You know the sort of thing.

Anyway, I saw an ad in *Buy and Sell* advertising a Sussex cockerel (let's call them cockerels and keep the innuendo to a minimum from here on), so I headed off in the car up to a small village in Tipperary to collect him. The woman I bought it from persuaded me that I should also buy a couple of bantam hens – she said they were great 'broody surrogates', which means they will quite happily sit on a load of eggs to hatch them even if the eggs are not their own. Bantams are like little miniature hens and they are sort of wild. I bought two – a cute little black one and a not-so-cute red one.

Mrs Kelly named the cockerel Roger, just in case there wasn't enough innuendo flying around the place. He's an impressive-looking chap – easily twice the size of any of the hens. He's white, with a massive shock of black and purple feathers on his tail, which has the effect of making him look a bit like a peacock. He struts rather than walks, and I always imagine that there should be disco music playing in the background whenever he's around. In fairness to him, he also lives up to his name – it wasn't long before he had got into his stride and started rogering all the hens daily – even poor old crippled Dora, which there must be some sort of a law against. The actual process itself is fairly gruesome to watch – there's nothing consensual about their coupling and in any other walk of life Roger would be doing multiple five- to ten-year stretches in Mountjoy. He sneaks up behind them while they are innocently pecking away in the grass and therefore have their nether regions up in the air in a disturbingly vulnerable manner. He jumps up on top of them, pinning their wings down with his claws and their heads down with his beak. He does his stuff and then jumps off again. It's all over in seconds (Roger is fundamentally selfish in the bedroom), the hen shakes herself off and goes on about her business. When I see the hens pecking in the grass with their butts in the air I feel like shouting: PUT YOUR ASS DOWN, YOU'RE IN DANGER!

After a few weeks he had his new harem in good order. He would give an impressive *Cockadoodledooooooooooooooo* and they would all fall in line. The crowing was cute to start with. Mrs Kelly especially liked it as it gave the place a distinctly farmyard feel. But I don't find it so cute when it happens at first light. When he arrived, first light was about 6.00am, but as the summer went on it got earlier and earlier. Mrs Kelly and I argue about this, but it's hard to have a proper argument about an animal that's called after male genitalia without it degenerating into convulsions of laughter. I'll say, for example: 'That damn cock had me awake at five this morning' and she'll just explode into laughter. She says she loves being woken up at dawn, since she knows she can drift back off to sleep for an hour or two before she has to get up. I lie in bed considering plans to turn Roger into a nice *coq au vin*.

Poor old Dora recovered somewhat from her initial rendezvous with the greyhound but met a grisly end a few weeks later. We went out for some grub one night and the hens were still out because it was still daylight. When we got home I locked the door of the hen-house but didn't have a torch with me, so I didn't look inside to make sure they were all there. Next morning I looked out the bedroom window and saw a familiar bundle of feathers in the middle of the lawn. When I went out, there was no sign of Dora. The little mound of feathers was all that remained of her. It was like she had spontaneously combusted. One of the other hens was also missing, but eventually came out from the ditch – she had obviously escaped somehow. She spent the next few days in a sort of daze, clearly traumatised by her ordeal. She would stand in the one spot with her head tucked in beneath her wing and if you touched her she would just fall over. Really strange. Whatever it was that got into the garden (probably a fox this time), it must have targeted Dora as easy pickings since the poor little thing couldn't run away. I felt so bad for her.

I decided it was time to take the predator threat a bit more seriously. You can get very complacent with hens wandering around on your lawn, thinking they are safe as houses. We didn't want to keep them at all if they had to be cooped up all the time in the run, so I went down to the farmers' co-op and got some fence posts and chicken wire and fenced off a small paddock for them. This has the advantage of allowing us to keep them confined in the morning until they all lay an egg, so that we don't have to go foraging for eggs in the ditches. It also means they still have quite a bit of space if we're going away for a few hours and want to put them in. I saw a design on TV for a hen-house on stilts which I also replicated. The idea is that the hens can climb up to safety if a predator attacks. I got a cheap, small dog kennel and put it up about a metre off the ground on four fence posts (more fence posts – the lady down at the co-op must think I have about a hundred acres of land with all my fencing projects!). I put two perches into it, attached two nesting boxes to the side and then made a rickety old ladder using some branches and twigs – it looks completely naff, but that's intentional. A hen can, with difficulty, get up the ladder, but a fox, hopefully, can't. The hens took this new contraption to their hearts – they definitely seemed to like the bit of height for laying eggs and night-time perching. And in no time at all they were fluttering up the ladder. They didn't seem to like coming *down* the ladder as much, though, and, instead of going from rung to rung, they did a crazy, semi-suicidal leap from the top.

And hens being hens, just when you think you've finally nailed it and discovered the ultimate hen dwelling, they will find some way to make you second-guess yourself. One of them decided that she didn't like laying in the new house and started flying out over the fence every day to get to her favourite spot in the ditch. Every morning at about nine o'clock she would start an elaborate ritual: get up on top of their house, fly into the vegetable patch which is also fenced off, then fly over *that* fence and totter off to the same little spot in a ditch and lay her egg. Instinct is a remarkable thing.

Once she'd laid her egg she would spend the rest of the day trying to figure out how to get back. For some reason, she seemed to be full of resourcefulness, skill and aerial agility *before* laying the egg but afterwards she became a complete bimbo, spending the entire day pacing up and down by the fence looking for a way back to her buddies. A few times I caught her in the act and put her back in the nesting box to try and encourage her to lay there, but no, she would just start out on her adventure almost as soon as I had my back turned. I tried clipping her wings, which is a perfectly harmless procedure where you snip the first five or six feathers on one side – it apparently makes the hen unbalanced in flight so they just don't bother. This didn't make any difference whatsoever. It used to really piss me off seeing her out and about, thwarting me at every turn, but now I actually think she's my favourite – you've got to admire her tenacity.

Since Roger's arrival all the eggs that we have are fertilised eggs and so all they need is for a hen to go 'broody' and sit on them for a few weeks for them to hatch as chicks. You can still eat the eggs, incidentally, if you don't have a broody hen to put them under, and some people think that a fertilised egg is better for you, but I find that hard to believe. You can put between seven and twelve eggs underneath a sitting hen, so we stocked up about ten eggs, waiting for the day that one of them would go broody. Nothing happened. To try and move things along a bit we left the ten eggs out in one of the houses. But it didn't work and after a few days I gave up and gave the eggs to the dogs as a high-protein breakfast.

Thwarted, we stopped stockpiling the eggs. And then, sod's law, one day the little bantam, for no apparent reason, decided to go broody. You know it has happened because instead of coming out of the house after laying her egg, she just sits on it. I noticed one day that she was in there for about four or five hours. But there were only two measly eggs under her. Thankfully I had a couple of eggs left in the kitchen which I put beside her too and the next day when the hens laid I put their eggs beside her as well. A broody hen gets very irritable if you go anywhere near her nest and she rewarded me for sticking my hand into her house by giving me a right peck. In the end we had six eggs under her.

Seeing a hen go broody is an incredible thing. They stay sitting for about three weeks, getting up off the eggs once a day to get some food and water and take a pooh. Some days we had to sort of hoosh her off the eggs – with great difficulty and much flapping and squawking – to force her to get some food. The natural movement of the hen while she is sitting on the eggs keeps them rotating ever so slightly, which keeps them at a constant temperature. We moved her into the old house by herself, which meant that the chicks wouldn't be in danger from the other hens when they were born.

Nature is amazing, really. The whole process is an incredible endurance test for the hen, a marathon of sorts, sitting there more or less 24/7 for three whole weeks with little access to food or water. Even more mind-blowing is that in just over twenty days the egg that you normally eat for your breakfast will magically transform itself into a fluffy yellow chick. How can the little egg which we love to poach, fry, boil or scramble possibly contain all the elements of a chick? What is it about some strange hen sitting on top of it that makes it embark on this extraordinary transformation? I think every child should at some point in their childhood get a chance to see a hen going broody and then see the little chicks three weeks later – it really is an enlightening experience. Even for us cynical adults, it was incredibly exciting when we heard the first, distinctive *Cheep cheep* from inside the hen-house. I opened up the door and the hen was still sitting there – no sign of any chicks, but I could hear the chirping, louder now. And then all of a sudden, I saw a little yellow head pop out from under the hen's wing, and then another! In the end, only three of the six eggs hatched, which was disappointing – later I read somewhere that you should take any chicks that hatch away immediately because sometimes a hen will get up off the eggs to tend to the hatched chicks, leaving the other eggs get cold and the chicks inside them to die. Maybe that's what happened. Anyway, as the old adage goes – you shouldn't count your chickens until they hatch. A day later I noticed there were only two chicks left. I checked the hen-house for a dead chick and couldn't find it anywhere, which was very disturbing. We can only assume that either (a) Mom ate it or (b) she threw the dead chick out of the house and something else ate it. Neither of these two possibilities is particularly nice to think about.

The two remaining chicks really were the cutest little things imaginable. They spent their first few days hiding beneath the hen's wing, peeking out every now and then to see what was going on. They developed amazingly quickly and were remarkably robust within two days. Mom was parading them out in the run and showing them the ropes – scratching in the ground, feeding and drinking. We gave them a special chick crumb for the first few weeks and put water in a shallow bowl for them – anything deeper and they'd go in for a swim and never come out.

The bantam hen looked after them for a little over a month and after that we returned her to the flock and left them in the run alone (we didn't want Roger getting his mitts on them). Chickens go through an awkward phase between cute chick and adult bird – as children move from being cute kids to handsome adults via our teenage years, when we look like we have the plague. After about six weeks the chicks start to lose their fluffy yellow feathers and they get one or two horrible white ones. Then, gradually, all their lovely little feathers are gone and they have nothing but horrible white ones. Eventually at about eight to ten weeks, once they have all their new feathers they turn into nice-looking chickens. I have no idea what sex they were, incidentally – determining the sex of a chicken seems to be like a dark art. I've read up on it and I am still none the wiser. I suspected they were male because I saw them squaring up to each other, but who knows? Anyway, if you are rearing them for the table it doesn't really matter what sex they are.

After some careful consideration, I decided it would be a good rite of passage for us (well, for me anyway) to go through the whole process, cradle to grave – kill the two chickens, pluck and gut 'em, and have ourselves a feast. It's a big decision, to kill an animal for food, and not one to be taken lightly, but I'm convinced that there is something profoundly hypocritical about being a meat-eater and then getting all squeamish about doing the killing yourself. Up until then the closest I came to killing was helping my father-in-law kill turkeys at Christmas. They kept six of them and Mrs Kelly and I went down to help with the killin' and the pluckin' a few days before Christmas on the promise that we could take one away with us. A turkey is a big old bird to kill and you need to know what you're doing to get it over quickly and efficiently and minimise the suffering for the animal. We grabbed them one at a time from their house, hung them by the feet in the barn, and the father-in-law pulled down on the neck, dispatching them to the nether-world with a quick yank. It was all over in seconds, after some reflexive flapping of the wings. It's not a particularly pleasant thing to see, but if you like turkey, it's something that you *should* see – if only to understand what the bird that you are eating has gone through on its journey to your plate.

I'm not ashamed to say that ultimately I would love to get to a point where I would be sufficiently skilled in terms of rearing, killing and processing chickens to be able to do it regularly – regular enough that all the chickens we eat would be birds that I have reared and killed myself. I've been amazed by people's reactions when they look in at the chicks and I tell them that they are being reared for the table – a mixture of sympathy (for the chicks) and disgust at the brutality of it all. We have become so detached from the realities of where our meat comes from that it's almost like we don't want to hear the reality at all – and when it's put in front of our noses, we recoil in horror. You pick up a nice, packaged, free-range chicken in the supermarket and it's easy to banish to the back of your mind the realities of that bird's life and death. Eating and cooking have become pretty sexy pastimes these days – sexy cookery shows with trendy young chefs licking cream off a spoon or caressing a large melon for our sensual pleasure. Feeling suitably aroused, we dash off to the supermarket in a pre-coital frenzy, loading the basket with the ingredients for the sexy recipe and dreaming of the sexy meal that we are going to have – and the sex that will surely follow. In that environment it's hardly surprising that we manage to forget completely about the phenomenally un-sexy reality for the poor animal that gave up its life for the meal.

Somewhere along the road to the land of happy convenience where we now live, we have managed to put serious distance between the act of eating meat and its brutal realities. It was not always thus. Thirty or forty years ago, when many people kept a few chickens in their garden, someone would go out on a Saturday, grab a chicken, wring its neck, and on Sunday the family would sit down to a rare, tremendous treat. There was the feeling that the chicken gave its life for a good cause and everyone who ate it was thankful for that. But when we run into the local shop and grab a chicken-caesar wrap or some chicken goujons, we don't have the same appreciation for these things. We don't even think about it. Perhaps that's not such a bad thing, I don't know. It's a difficult issue. Having done the deed myself, part of me thinks we should just say thank God we don't have to think about these things, and move on. But another part of me thinks that it would probably do us no harm to take a moment to think about it every now and then. That maybe it has all become too sexy, too convenient, too easy, too cheap.

As I've said, an organic chicken takes about a hundred days to rear, but our two still looked very scrawny after the hundred days, so we let them be. Apparently, you are supposed to try and imagine the chicken *sans* feathers and lift it up and feel the breast to see if it's ready. I did both of these things, but I still was uncertain as to their meal-worthiness. At about 120 days I decided to go ahead, regardless, with one of them and, if that one was small, I'd hold off on the other. You can't keep chickens for the table indefinitely – beyond four or five months both cocks and hens become pretty unpalatable for anything other than a soup or casserole (cocks, apparently, becoming tougher first). The mother-in-law gave me a loan of a very handy little contraption for the execution – it's a little metal frame that you put the chicken's feet in, hang them upside-down, and then you wring their necks and pluck them while they are hanging (*à la* turkeys at Christmas). It's a grim-looking device, rather like a medieval instrument of torture, especially when it's hanging from a piece of rope from the rafters in the garage.

Each time I fed the chickens I was reminded of the deed that lay ahead. My sister kept trying to talk me into letting them live, as if I was a governor of some gung-ho, electric-chair US state being asked to pardon a prisoner on death row. 'Why don't you just leave them be?' she would implore, any time she came around. In some ways it was tempting, but I also felt it would be completely ridiculous to do so, unless I was never going to eat meat again. I had a dream one night that I was looking out the bedroom window and there was a large gang of anti-death penalty protestors walking around the coop carrying placards saying: 'Don't Fry the Dixie Chicks' and 'Meat Is Murder'. There were pro-death penalty people there too with their placards, just to balance things off and breathless TV reporters were yelling to TV cameras, sending updates on my emotional hand-wringing around the globe. Such was the agonising that went on in my head over the issue!

As the hundred-day milestone came and went, I kept putting off the inevitable: don't have time today, will do it tomorrow. Too busy this week, will think about it next week. I was sitting at my desk one morning, writing, and I suddenly decided, for no particular reason, today's the day. Out I went. It was a particularly sunny day and the two chickens were sitting in their run, sunbathing. I had a brief pang of guilt, thinking about the miserable weather they'd endured that summer and wondering whether I should let them enjoy the sunshine. But then I gave myself a good talking-to.

I didn't put a lot of thought into which one was for the chop first and just grabbed the first one that came to hand – there was the usual kerfuffle, which there always is when you try to catch a bird. Once I had him under my arm he calmed down and didn't make a sound, even when I brought him into the garage and hung him upside down from the torture device. He was completely still – I caught him underneath his head with two fingers and pulled down, he flapped around for a while, maybe thirty seconds, and then was still. That was it. I had been worried that either (a) the head would fall off, which can apparently happen, or (b) I wouldn't be able to kill him at all and it would turn into an amateurish bloodbath with me beating him over the head with a mallet to finish the job. But, thankfully, neither of those things happened and in the end it was relatively quick and (I hope) painless. Strangely, after all my agonising, that was the easy part.

Plucking a chicken is a particular skill and not one that I know much about. I did remember from helping the in-laws with the turkeys that it has to be done quickly while the bird is still warm, otherwise the feathers are very tough to pull. I also remembered reading something about dunking them in hot water to soften the feathers, but I was pretty sure we hadn't done that when we killed the turkeys so I decided to proceed without it. It's an arduous sort of a job to do when you're doing it alone and don't really know what you're about. It took me well over an hour. The bird was still hanging upside-down at about chest height, so it was relatively comfortable work. While I was doing it, I was thinking: Bloody hell, I can't believe all this work has to go into one meal!

It's very odd when you hold the bird and the skin is still warm – but slowly, as the feathers come away, it starts to look more like a chicken in the supermarket sense and less like a living, breathing, pecking chicken. During that transition my heart rate gradually slowed down and my general feeling of unease was replaced with a sense of purpose and excitement about the meal ahead. The trick with plucking is to try not to tear the skin – sometimes when you pull on some feathers the skin tears and this is obviously not very desirable because it exposes the meat underneath. I imagine this wouldn't have happened if I had used the hot-water technique, but I still have to verify that. I had a plastic sheet underneath the bird to catch the feathers – another trick learned from the in-laws – it felt very *CSI Miami,* like I was sanitising the crime scene after a vicious strangling.

As the final feathers were plucked I was pretty despondent at the pathetic meal that was unfolding before my eyes – the poor thing looked absolutely tiny, more like a little rabbit than a chicken. Anyway, it was evidently too late to do anything about it now. I carried him inside to the kitchen, past the dog who was sitting on the deck sunning himself. When he saw me with a bald, dead chicken, he seemed to sense that something very important had happened. He's used to the hens being out and about in the garden and mainly he tolerates them, but I could see his eyes all wild and he must have been thinking: *What does this mean?! Has there been a fundamental shift in our relationship with those pesky birds??! Does this mean we are free to kill them?!*

My first attempt at gutting was pretty amateurish and very nightmarish. Basically, I used a very sharp knife to cut off the neck and head, then the feet – and then came the worst part: I stuck my hand in and pulled out the bird's innards. This was the only time during the whole process that I was really grossed out – it's absolutely revolting. Revolting smell.

Revolting feeling. The guts, gizzard, lungs and heart came out in one big, foul bundle and I had to stop myself from retching. But once you have that done, you're at the stage you would be at when you unwrap a chicken that you've bought in the shop. They say you should hang a bird for a day or two in a cool place, but given that it was really warm the day I killed it (which was probably a mistake) I decided that would be a bad idea. About two hours after I had wrung his neck, the chicken was plucked, gutted, jointed and sitting in the fridge, marinating in some olive oil, garlic and rosemary.

When Mrs Kelly came home from school, I told her I'd done the dirty deed. She was horrified, delighted and impressed in equal measure, but mainly she was happy that she hadn't been involved. That night we had the breasts with a salad and some potatoes – I suppose another benefit of leaving it a few days before eating it is that you feel a little more removed from the process of killing it – I have to admit it felt a little odd eating the breasts that night, although they were very tasty. We had a very nice casserole the following evening with the legs and then I made a stock from the carcass, which became a delectable carrot soup. And that, as they say, is that.

The thing that struck me most about the whole thing was what a poor return it was for all the effort. I mean, don't get me wrong, they were two nice meals and it was very good soup, but let's take a moment to think about the work that went into it. It took over twenty days for the chicks to hatch in the first place, another 120 days to rear them, an hour to pluck the bird and perhaps another hour to gut it and prepare it. It just seemed like we deserved more for our efforts. It probably would have helped if the bird had been a decent size, but basically the end result would have been the same – what you get from all that work is one or two meals. I'm not saying that it's not worth the hassle – in my opinion it's worth anything to know where the meat you are eating comes from. I suppose I'm saying that I was just flabbergasted at how much work goes into producing such a small amount of meat.

In hindsight, I can say that I didn't enjoy the kill or revel in the goriness of it, which is as it should be. It's a life, after all, and we shouldn't take any pleasure in extinguishing it. In fact, I found the whole thing pretty disgusting – especially the gutting – and I will own up to some very strange dreams (OK, nightmares) that night, even stranger than my 'Governor Kelly wrestles with his conscience' one before I killed them. I think you could probably get used to the gore and guts eventually, though I'm inclined to think it would be easier to buy a large batch of day-old chicks and kill them off all at the one time (as the crafty and experienced mother-in-law does, of course), with a little help drafted in to speed things along. That way, you would get all the hardship over in one day, fill the freezer with maybe half a year's supply of chicken and then let the bloodlust subside for a while.

Nonetheless, there was an element of satisfaction when it was all done. I was proud of myself that I had gone through it. That I had taken responsibility for rearing an animal from start to finish, then taken responsibility for killing it with my own hands, then, a few hours later, eaten it with a salad and some potatoes. There was something particularly grounding about the fact that I 'knew' the animal – if you can ever really *know* a chicken – before I killed it. I suppose it was an epiphany of sorts for me, the moment that I overcame my own hang-ups about the strange relationship we all have with the animals we kill for meat. It was the first time I could reconcile going to all that trouble to ensure an animal is happy in its house, secure, well-fed and contented, and then deciding to end its life. As a result, I don't think I could ever pick up a breast of chicken in the supermarket with the same nonchalance. Once you've killed a chicken with your own hands, how could you?

Chapter 8

THE DOWNSIZER'S DILEMMA

Every Downsizer arrives at a crossroads of sorts at some point in the journey towards austerity when they wonder whether they have the heart to be a Downsizer at all. Since we moved to Waterford I've often wondered whether our downsizing is the result of a genuine desire to change the way we live – to downshift, opt out of blatant consumerism, root out all unnecessary expenditure and live a simpler, more frugal existence, or is it just a tactical, temporary expediency, a cunning ploy to help us survive some undoubtedly lean years? What would happen, for example, if an old aunt suddenly passed on and decided to leave me a grand fortune? What then? Would I still be interested in putting in the hours in the garden, sweating and heaving, up to my neck in muck, lugging wheelbarrows around the place, all for the grand reward of some spuds and a few turnips? Would I still be happy to clean up mountains of hen-shit in exchange for a couple of eggs, or wring a chicken's neck and pluck a million feathers from its body for two measly meals?

This is what I call the Downsizer's Dilemma. Internally, in my own psyche, the constant battle of wills rages between my penny-pinching, green-fingered Environmentalist and my fun-loving Capitalist, with some part of me in the middle keeping score. On the one hand, I'm happy that we've learned to be more careful with money and can live on a small income. On the other hand, I'm still a person who likes *stuff* and appreciates bling. Metaphorically speaking, it's like going to buy a car and discovering that your head wants a Prius, while your heart wants a Hummer. Part of me is happy that we spend our holidays in Ireland and I feel pretty good about the fact that we've saved money and been kinder to the environment by not taking unnecessary flights. But there's a part of me that wants to be John Travolta so that I can collect private jets just for fun, and fly off at the drop of a hat for a short city break or to my luxury villa in the Azores. Can these two fundamentally opposing ways of life live side-by-side? Can they ever be friends?

I've come to the conclusion that to a certain extent they can, but never completely, and sometimes the row ain't pretty. I'm starting to think that essentially I am a Downsizer trapped in the body of a Capitalist, or maybe it's vice versa. Sometimes the Downsizer is in the ascendancy and at other times the Capitalist holds sway. This explains why some of my efforts to downsize have been a phenomenal success, while others have been ... well, basically, a pathetic failure. It explains why so many of the 'Giving Up' experiments – the mobile phone, TV, driving – were temporary as opposed to longterm behavioural changes. The Downsizer liked the idea of opting out – but the Capitalist was delighted it was only for a week or a few weeks and loved getting back to normal. It explains why I get so excited about little hare-brained projects that might help save some money – joining the library so that I'll stop buying books, fashioning logs for the fire out of soggy old newspapers, making homemade deodorant out of oils and beeswax – but also why, eventually, I get fed up with the hassle and revert to type.

In a pretty typical fit of pique one week, the Downsizer took a scissors to our credit card and cut it right down the middle, arguing, with considerable logic it has to be said, that a credit card had no place in our brave new world of reduced earnings. But quietly, and with little fanfare, a few months later the Capitalist called the credit-card company, apologised on the Downsizer's behalf and asked them to send us out a new one. Just for emergencies, you understand. The Downsizer called up Sky one week and asked them to cut off our Sky Sports subscription. The Capitalist smiled one of his wry little smiles and said that he was pleased that it was just Sky Sports that was gone (it could have been worse, it could have been all the Sky channels, or, hell's bells, the TV itself) – and anyway, he has his eye on the Ryder Cup in September as perhaps a suitable juncture to get Sky Sports switched back on.

The ultimate front line in the battle between the two is the debate over whether or not we should hold onto my car, a 5-series BMW that I bought when I worked in sales. This is the most vicious, bitter and antagonistic issue on the table in front of my two personas. It is the line in the sand and an issue that neither of them is willing to concede ground on. In the early days of my journalism career, when I was getting paid about as much as a sixteen-year-old temp worker in a fast-food joint – worrying times indeed for the Capitalist – my BMW was like a mask that he could put on to pretend we were still an affluent person. My car is actually a 2-litre, '99 reg behemoth that has 150,000 miles on the clock and is well past its prime, but, hey, it's still a BMW – a big, bad, brash statement of affluence and almost certainly not the car of a struggling hack. The Downsizer tells the following (ancient) joke to show his feelings about BMWs: What's the difference between a BMW and a hedgehog? A BMW has its pricks on the inside. The Capitalist doesn't find that funny.

I know that I should sell my BMW and replace it with something more in keeping with the downsizing ethos – maybe a little battered-up Fiesta with roll-down windows and a rust problem, or a trendy scooter. But the part of my personality that enjoys nice things, that wants to be *seen*, at least, to be doing well, has won through thus far and so the tank is still parked in the driveway, much to the Downsizer's chagrin. I'm not particularly comfortable with the fact that I am this shallow, but what can you do? I wouldn't mind, but the car doesn't even get used all that much anymore – it sits forlornly outside the house for weeks on end, practically beseeching me to take it for a spin (occasionally I indulge it with a trip to town for the shopping). But the Capitalist doesn't really care whether it actually gets driven at all – so long as it's *there*, he's happy.

It is not surprising, perhaps, that the car situation is the most pressing of all amongst the myriad of issues in the Downsizer's Dilemma. For the car, more than any other item that we own, is the one we use most to show the world how well we are doing (especially us males). It's a status symbol on wheels. Which is a pity, because the massive loans that we need to pay for our new set of wheels have the potential to lock us into jobs that we don't enjoy and stresses and strains that we shouldn't have to endure. I bought this car in cash with a bumper commission cheque and some of the profits from the sale of our house in Gorey. I am sure that financial experts will tell you that this not the right thing to do at all, what with the cost of borrowing being so low etc, but actually it's quite liberating because it frees you from monthly repayments. The fact that neither of us has a car loan is an important factor in allowing us to survive on drastically reduced earnings.

Take a look around at all the top of the range, brand new cars that are on the roads – the overwhelming majority of them are funded by car loans. A Range Rover Sport with a 4.4 litre engine, for example, will cost between €107,000 and €147,000. The repayments on a car loan that size will be approximately €4,000 per month over twenty-four months. Imagine how much money you need to earn to be able to afford a statement of affluence that size? Most people pay about €300 to €500 on a car loan each month. Imagine if they didn't have that expenditure in their lives? They would either have more money each month to play with or (and this is where it gets really interesting) they could afford to earn less each month and not have any impact on their standard of living. That dream job, which doesn't pay as much, suddenly looks more attainable.

If I was driving a brand-new BMW, we could agree that the Capitalist had won the argument, and leave it at that. But because it's an *old* BMW the Downsizer feels he can potentially sow some discord between the Capitalist and me – he senses that, in fact, the Capitalist is not particularly happy with the behemoth. Where a new BMW is a statement of affluence, a second-hand, old BMW is a statement of affluence on a budget – a yellow-pack status symbol, which is the very worst kind. At least someone driving a brand new luxury car has either the financial wherewithal or a sympathetic bank manager to back up their sizeable ego. In some ways, because it's an old car I have the worst of both worlds.

I was at a conference in Dublin recently and because I arrived late, I parked my car in front of another car just outside the front door. Just before the mid-morning interval one of the speakers stood up at the podium and said: 'Will the owner of the following car registration please remove their car as it's causing an obstruction' – and proceeded to call out mine. Mortified, I stood up, and made my way to the door with the whole room watching me. I realised that my embarrassment was due to the fact that I thought everyone in the room was judging me for driving a car that's nearly ten years old. Who does that reflect badly upon? Me? Or Irish society in general? It's me, isn't it?

I reckon that if you are happy to drive around in some battered-up piece of shit you can congratulate yourself on having pretty much eliminated pride from your personality profile – and I don't mean healthy, wholesome pride, like the pride you feel when you get a promotion or something, I mean, dirty, choking, clawing, seven-deadly-sins, you're-going-to-hell pride. Clearly, with the ongoing Downsizer's Dilemma going on over my car, I don't fit into that category at all.

Now, Mrs Kelly on the other hand, drives around in what must be the oldest, crappiest car in Ireland. It's a 1994 Toyota Starlet, which I lovingly call The Jalopy. Most people know that Toyotas are durable cars and that's why they fetch such good prices when sold second-hand – but her Starletti takes that durability to the extreme. It is like the world's oldest person, it just refuses to die. It has passed its NCT every two years with flying colours and has never given a day's trouble since she bought it. It was her first car and although most people her age have had four or five models, she's had just the one. She refuses to get rid of it. It's almost a stubbornness at this stage, although clearly she's very fond of it – and she argues that since it still performs its basic function of getting her from A to B with such obvious aplomb there would really be only one reason to change it: so as not to be *seen* in a twelve-year-old car. And she doesn't go in for stuff like that. Damn her.

A couple of months back we decided it might be an idea to sell the behemoth and get a nice, sensible, mid-sized car which Mrs Kelly could use for going to work and, given that I hardly drive at all anymore, I would take over The Jalopy. This seemed like exactly the kind of reasonable accommodation that Downsizers should come to. The Capitalist was, of course, horrified at the idea and slyly suggested that we test whether my fragile ego was up to it by persuading Mrs Kelly to swap cars for a week – I drove the Starletti around a bit while she travelled to work each day in understated German luxury. It wasn't a happy arrangement. She found the behemoth too awkward and big for her liking, and I was alarmed to discover that I am, in fact, a complete and total car snob.

Let me tell you this for starters. You feel really, really vulnerable in an old, small car. Modern small cars are not really small at all – they are way bigger, wider, taller and longer than they used to be. Car manufacturers tell us that they are responding to lifestyle changes with these so-called superminis – which is a nice way of telling us that we are all fatter than we used to be and carry more crap around with us. I was in traffic one day and a woman pulled up beside me driving a Toyota Verso, which I think is more or less the replacement of the Starletti, and she may as well have been driving an articulated lorry, the way she was towering over me. (Incidentally, she looked down her nose at me and I could clearly see she was thinking: that guy has a tiny penis.) Apart from the constant feeling of terror that any minute I was going to be squished like a little ant by other road users, I actually found very little difference behind the wheel between a Starletti and the BMW. A car is a car is a car.

The problem I had was not the driving experience itself, it was worrying about how I would be *perceived* driving around in The Jalopy. I was coming out of a shop in the village with someone and I actually held back and let them go on rather than approach The Jalopy which was parked outside the shop.

When I finally sat in the car, satisfied that nobody had seen me, I was disgusted at my own shallowness. My reaction confirmed what I knew deep down all along – I hadn't bought the BMW because it would be safer or because I love its handling and precision engineering. I had bought it to make a statement. We all buy into status symbols on some level and we are willing to spend lots of money (ours or a bank's) to acquire them. Our clothes, where we live, our house, our car – all part of a carefully constructed message about who we are. That's the grease that oils modern Ireland, folks.

Anyway, having stood by, helpless, as the Capitalist scuppered plans for me to take over the Starletti, my next cunning plan was that I would sell the behemoth and buy a scooter. I figured it would be grand for heading up and down to the village for the paper and wouldn't cost much to run. I had driven one when we were on holidays in Croatia about five years ago but Mrs Kelly ruined the experience from the pillion passenger's seat, screaming at me to stop going so fast (I was doing about 5kph) and squeezing so hard around my waist that I could hardly breathe. But that didn't put me off. I had visions of myself as Dunmore's answer to Jamie Oliver, tearing around the village on my Vespa, stopping to buy fresh fish at the market and shouting '*Ciao*!' and '*Pukka*!' at everyone I met along the way. Then, one day, as if fate was knocking on my very door, I saw an ad in the local paper advertising a secondhand Piaggio scooter for sale for €150! Ignoring the old adage – if it sounds too good to be true it definitely is – I jumped into the behemoth and went to see it.

The man selling it told me that his son had gone to Australia for a year or two and he was selling it because it was taking up too much space in the garage. 'He won't be too happy when he comes home,' I mused. 'Ah sure, he'll be grand,' he insisted. 'He can buy another one.' 'Why so cheap?' I asked suspiciously. He told me he just wanted to get rid of it quickly and it was pretty old anyway. He took me into the garage and there it was. My heart sank. It wasn't the scooter I'd expected – I was thinking of a sort of retro *Roman Holiday* job, but this was just like the one I had rented in Croatia, all garish colours and go-faster stripes. But it was too good an offer to dismiss it on aesthetic grounds alone, so when he asked me if I wanted to take it for a spin I said, 'Sure, why not?' When I sat on, I realised I couldn't remember how they worked, but, like most good men would, I ignored this obvious limitation and carried on.

'Have you driven one before?' the guy asked, looking slightly alarmed as I struggled to get the helmet on. 'Oh yeah, loads of times. But it's been a few years.' He gave me a quick demo, and after a couple of wobbles, off I went down his driveway. The guy's house was up a long cul-de-sac, so, thankfully, there were no cars about. I went up and down the road a few times slowly, to get my bearings. The first thing that struck me was that scooters are really, really LOUD. They make this hair-drier-fed-through-a-microphone type noise which is intensely irritating. I imagined myself driving through the village and everyone looking up at me, tut-tutting at the noise pollution, my friendly '*Ciao*' drowned out in the din of hairdryer-engine noise. *This is fecking mad, what the hell are we doing here*, moaned the Capitalist as I approached the end of the cul-de-sac. *We're thirty-three years old, not sixteen. It's bad enough that you make us use a pre-paid phone. We're not buying this piece of shit! Come on, let's go for a latte.*

My mind finally made up, I went to turn around and realised I was moving too quickly – my instinct was to brake, but instead of using the brake I actually grabbed tighter on the accelerator and as I turned, the bike sped up, mounted a pavement and crashed into a wall. I got up and removed a sod of grass which was lodged between the front wheel and the mudguard. Apart from that, there was no damage. I wheeled it up the road, told the guy I would think about it and drove off. The Downsizer sat in the back and sulked all the way home.

Chapter 9

PIGS IN SPACE

'Half the dogs in America will receive Christmas presents this year, yet few of us pause to consider the miserable life of the pig – an animal easily as intelligent as a dog – that becomes the Christmas ham.'

The New York Times 11/10/02

We are exceptionally lucky to have as our neighbours a young Amish couple who are further on the road towards self-sufficiency than we are – their success acts as a kind of motivator for us when we slack off, which we invariably do. Last summer they asked us to mind their hens for a few days while they were away, which was fun because their hens are an unusual breed called Brahma that make our hens look positively dowdy. They have two hens and an impressive multi-coloured cockerel, and all three have heavily feathered legs right down to the claw, so it sort of looks like they're wearing leg warmers or flares. The cockerel is called Charlie – he's a macho sort, of course, but is somewhat let down by his vaguely pathetic crowing, which starts well but then tapers off to a death-rattle type croak. I always smile to myself when I hear that off in the distance. Anyway, as a thank-you for minding Charlie and his lady-friends, they called over with a bag of frozen pork chops from their own pigs. The nicest present you could possibly get.

I'd been going on for quite a while about getting some pigs, but the decision to take the leap into porcine husbandry is not to be taken lightly. Keeping hens and keeping pigs are worlds apart. Pigs are livestock. They need tags from the Department of Agriculture. To get them slaughtered you need a herd number, which, in the eyes of bureaucracy, makes you a herd owner. You even have to have a Department of Agriculture inspection. Hens are just hens. You are not likely to be in any way alarmed by a hen running at you; but if a fully grown pig runs at you, you are likely to be very alarmed indeed. If a hen reaches the end of its natural usefulness you can, theoretically, pick it up and end its life with your own bare hands. That's the idea, at any rate. With a pig, there are abattoirs involved. There's slaughtering. Butchery. Keeping pigs is, dare I say it, *farming*.

Whenever I think about pigs I always think about the sequence in *The Simpsons* where Lisa decides to become vegetarian. Homer wonders will she have to give up all meats, eg bacon, ham, pork chops, until she informs him that these all come from the same animal. Wow! he thinks, that must be a wonderful, magical animal.

It is. In Ireland we consume nearly 40kg of pork each per annum. Ham, pork chops, roast pork, pork steak, salami, pepperoni and, of course, the most venerable of all dishes – the jumbo breakfast roll with sausages, bacon and pudding. Yum yum. Pork typically represents 40 percent of the average person's meat intake. A wonderful, magical animal indeed. My favourite pork product of all is the humble sausage. If I was given the choice of one food type to eat for the rest of my life, I'd opt for sausages. It would probably be a short life, but hey, what the hell! I love sausages, so it would be an enjoyable one at least. The old saying goes (attributed to Otto Von Bismarck, I think) that there are two things one should never see being made: laws and sausages. And you do have to wonder, sometimes, what the hell is *in* them? Curiosity is a terrible thing and I, for one, am mightily sorry that I ever bothered to find out. Are you ready?

The recipe for most cheap sausages is typically: 30% pork fat, 20% recovered meat, 30% rusk (basically a filler which helps the sausage to absorb water), 15% water and 5% assorted e-numbers – flavourings, sugar, flavour enhancer, preservatives and colours. You're probably thinking, as I was, here we go again, 'recovered' meat? Remember the chickens when they had passed their use? The official definition of recovered meat states that it can include skin, rind, gristle and bone. Yuk. Basically, your sausage is made of 30% pork fat and very little else of any consequence. Thankfully, there are alternatives starting to appear on our supermarket shelves and the premium sausage market is growing at approximately 25% per year. You just need to know what to look for. Flip the product over and have a look at the ingredients list. A decent sausage should include at least 70% pork meat, ie it should say '70% pork', not 'pork fat'. The sausages will be better still if the meat is a prime cut – preferably pork belly or shoulder.

You don't want skin, rind, gristle, bone or testicles in there. Steer clear of any products that mention *recovered* meat.

Premium sausages are expensive – you can pay up to €4 for a packet of six, which is a lot of money, especially when you consider that you can get various own-brand sausages for as little as €1. But it's important to remember why the premium ones are called premium – there's better stuff in them. We all have a tendency to take the lowest imaginable price for a product and establish it as the benchmark by which we measure all competing products, even when the product in question is the nastiest shite imaginable. The point is not that these premium sausages are dear – it's that the other ones are too cheap. Those own-brand sausages cost about the same as a tin of dog food. Come to think of it, if there is recovered meat in your sausages, the dog food would probably be better for you.

Recent years have seen massive interest in free-range and organic meat, particularly chicken, beef and lamb, but oddly, not pork. When's the last time you saw free-range pork in your supermarket? The answer most probably is that you never have. To understand this anomaly we firstly need to understand how the industry is set up. At a time when our appetite for pork has exploded (approximately three million pigs slaughtered per annum), the number of pig farmers in Ireland has fallen from over 60,000 in 1970 to approximately 500 today. Three million pigs from 500 farmers? That's one hell of a herd on each farm.

So there are fewer, bigger farms rearing the pigs. There are also fewer, bigger abattoirs killing them. The small, family-run, local abattoir has been all but eliminated from the market by the policies of successive governments and those meddlesome EU directives. Strict regulations impose massive costs on these small businesses but have done little to improve hygiene, eliminate disease or improve the lot of the pig (or the consumer). I know of one man in County Wexford who ran a small abattoir, which was by all accounts the cleanest and best imaginable – the owner invested serious money trying to keep his premises, processes and equipment up to the levels demanded by Department regulations, but in the end he closed down because he couldn't afford to stay in business. In 2007 the government announced a new €50m 'grant' scheme for large processors which the Associated Craft Butchers of Ireland called an 'attack on small and local processors'. In other words, the government were giving money to large business when they should have been giving money to the little guys.

With the small farmer and the local abattoir squeezed from the market, what we are left with is industrial scale pork production – that's good news for the consumer, in theory, because it means we can get our hands on seriously cheap pork (€1 for a pack of sausages). It's very bad news for the poor old pig. The overwhelming majority, and I mean almost 100 percent of commercial pigs in Ireland are reared indoors in cramped conditions, with no access to the outdoors. An intensively reared pig will never eat a blade of grass or a vegetable, or feel the sun on its back. Underfoot are concrete laths to prevent a build-up of pig manure in the sheds. Because they are kept indoors they will never get to indulge in their favourite piggy pastime – rooting, or 'rootling'. One of the most harrowing aspects of pig farming is the treatment of breeding sows – they are quite literally used as pig-producing machines. It is estimated that over half of them in this country are kept in stalls which are so narrow the pigs can't even turn around. These unfortunate sows produce six or seven litters in this environment and are then slaughtered. They are undoubtedly better off dead. This method of breeding will be phased out under EU legislation but not until 2013 for existing farms. The offspring of these unfortunate sows don't fare much better. Everything about their treatment during their short lives is about getting them fat, quickly. The quicker they get fat, the sooner they can be killed. The sooner they are killed, the quicker the producer can turn a profit on them.

It's plainly not right to treat an animal in this way, but most of us don't warm to the animal welfare issue when there's a pig involved. Unfortunately, the pig has something that makes its misery even more complete: intelligence. Pigs have lots of it – at least as much as your pet dog. Joanna Lumley did an experiment for the Compassion in World Farming organisation where she got a dog trainer to train both a pig and a dog to do some basic tricks – the pig learned the tricks quicker than the dog. Over in Penn State University in the US, scientists put a ball, a frisbee, and a dumb-bell in front of several pigs and were able to teach them to jump over, sit next to, or fetch any of the objects when asked. Amazingly, the animals could still distinguish between these objects three years later. In other words, pigs have really good memories. Another study suggests that pigs have the same cognitive abilities as a three-year-old human child. We get all misty eyed at the mere mention of someone harming a dog, and rightly so. Strangely, we have no such concerns about the welfare of the animal that gave its life to provide the contents of our breakfast roll.

The success of the entire industry is predicated on the animal welfare issue being kept well hidden from us and they do that by throwing a large cloak of regulation over it. Since we got our own pigs we've been put on the Department of Agriculture's mailing list, so I get regular letters and parcels in the post with information on how I should treat my pigs. They tell me the minimum space I have to allocate per animal – it's derisory. They tell me what I can feed them – it's depressing. If the animal welfare aspect doesn't ding your dong, then think about food quality. The pork that's available to us in the supermarket is, let's be kind and call it AVERAGE. These days when you cook a rasher on the grill you can see the grill through the rasher, and that's a bad sign. It's common knowledge that to make rashers look nice and meaty on the supermarket shelf, processors add water to the meat. When you cook the rashers all the water evaporates, leaving you with a shrivelled piece of cardboard instead of a piece of bacon. That white frothy residue that's left on the pan and on the rasher itself? That's the evaporating salty water.

It's not just what the pigs are fed and how they are reared that makes for shit meat. The environment the pigs are killed in also has an influence. A pig that is killed in a high-stress environment is one that is leaking adrenalin into its meat at the point of slaughter, and this has a disastrous impact on the taste of the meat. And the Ministry of Agriculture in Canada had this to say about the impact the stress of transport to the abattoir can have on the pig: 'the simple act of loading and short transport of say 1.5 hours can increase a pig's salivary cortisol levels by two to four times normal value ... such pre-slaughter stressors can have a negative impact on pork quality resulting in altered colour, moisture holding capacity, pH and toughness.' If the journey to the slaughter can have that kind of impact, imagine what the stress of a modern slaughter house will do to it? I'll talk a little later about the squealing our pigs did the day we brought them home. It's a noise I will never forget. Imagine a couple of thousand pigs being readied for slaughter. Imagine the noise they would make. Imagine the stress they would be under. Imagine what it's doing to your pork.

There are some abattoirs out there that understand these things and will take every effort to get the pigs relaxed before they're slaughtered. The slaughtering process is still pretty unpleasant, of course, but at least the stress for the poor animal is kept to a minimum. Lamentably, most of these small abattoirs have been put out of business by our government. You won't find these niceties in commercial piggeries – they don't have time. There are quotas and targets to be met. That's fair enough, in one sense, but there should be alternatives – there should be a mix of large-scale commercial operations *and* smaller local ventures to cater for the increasing number of people who want to do things differently.

I was at the farmer's market in Enniscorthy one Saturday and I met a chap there who was selling free-range pork. If you see anyone, *anywhere* selling free-range pork you really should buy it. These guys need our support. Anyway, this particular man was passionate about pigs and it wasn't long before we struck up a conversation about keeping them. I wrote down his number and when we had finally made up our minds to take the plunge and get a few of our own, I gave him a call. As luck would have it, he had four piglets left from a litter and they were about thirteen weeks old. They were a breed called Tamworth.

There is some debate about where the Tamworth comes from – whether it's an Irish or English pig – but they are generally considered to be quite unusual and quite an old, wild breed. A vet visited our two and told me that he had never seen anything like them before. 'They look like wild boar,' he said. My father-in-law, who had pigs on his farm when he was young, said they had 'common heads on them' (I'm still not sure what that means) and had 'an arse on them like a duck' – a reference to how lean their hams were. As you can see, the father-in-law has a wonderful turn of phrase.

They did look a little unusual. For a start, they were a rusty colour rather than pink and they had lots of hair. The Tamworth also has an unusually long snout which it uses to great effect for some serious rooting. I read somewhere that the Tamworth came top in a taste test run by some university in England, but the proof, as they say, would be in the pudding (mmmmm, pudding). They are also known for being slightly on the gregarious side and are probably best known as a breed for the heroic exploits of the 'Tamworth Two'. In 1998 a pair of pigs escaped while being unloaded from a lorry at an abattoir in Wiltshire in the UK. They swam across the river Avon and then ran into a dense forest where they holed up for a week. Their outlandish bid for freedom caused an international media sensation, with the world press dubbing them 'Butch and Sundance'. They were caught after a week and in the end the *Daily Mail* paid their owner an unspecified amount of money to save them from slaughter. Apparently they lived out their days getting old and fat at a Rare Breeds Centre in Kent. That's a pretty good analogy for our warped relationship with the animals we eat. We don't like to think about the plight of the millions of pigs slaughtered each year to satisfy our appetite for pork, but when we are forced to think about it, like in the case of the Tamworth Two, we start to root (excuse the pun) for the feisty underdog.

As it happened, my neighbour, whom I mentioned at the start of this chapter, was interested in getting two more pigs and since he couldn't source them from his usual supplier he said he would take two Tamworths. This was crucial since it meant we could go and collect them together – he had plenty of experience so I wouldn't look like a complete muppet. Also, I don't have a trailer, so I'm not sure how I would have got them home had it not been for him. Unfortunately, all that was left from the litter was three males and one female – it's generally accepted that meat from male pigs is less tasty than from females. One farmer I talked to once told me that boar-meat has a 'pissy taste'. Although I laughed at the time, I could sort of understand what he meant – once or twice I've got an acrid smell in the air while boiling ham, which apparently is the classic sign of boar meat. Anyway, when I mentioned this to my supplier, he was dismissive of the theory and said that boar meat only tastes bad when the pig is slaughtered while under stress as testosterone leaks into its body at the point of slaughter. He claimed that the meat would be fine so long as we killed it young (before it came to sexual maturity) and used a good abattoir. My neighbour agreed to take the two boars, leaving me with a boar and a sow. Being a complete novice in these things I enquired innocently whether it would be okay to keep a male and female pig together. This wasn't as stupid a question as I first thought – apparently pigs don't have the same hang-ups about incestuous relationships as we have so the fact that they were brother and sister wouldn't stop them getting it on when the time was right. My supplier was ada-mant that we would be killing the pigs before they reached maturity so this wouldn't be a factor.

We had a hard time trying to decide where to put the pigs. Some people I talked to told me they would be fine on con-crete, others said they needed grass. I wanted to leave them on grass, but I agonised over what sort of state the place would be in and whether our garden was too soggy (we have really wet soil). Having had pigs for some time now, and therefore considering myself quite the expert (not!), I can tell you this for nothing – I am so glad we decided to keep them on grass.

I'll come back to this issue later. After a tortuous debate, we finally decided to put them in an area down at the end of the garden – it's where we have the compost bin and I throw the grass cuttings and attic intruders, and is a complete no-go area, overgrown with ivy and weeds. There are two or three large trees in there too. I bought a small, battery-powered fencer and some more fencing posts (availing of the 'buy 500 or more fencing posts and get one free deal' down at the co-op). It's an area about 20m long and maybe 7m or 8m wide.

Next we turned our thoughts to housing. You can buy specially designed pig arks – there are even really fancy wooden ones similar to my homemade hen-house (only not so crap). But these are pricey. I thought about making one, but remembering the saga that was the hen-house construction I just didn't have the heart for it. Then my neighbour told me that he knew of someone who was throwing out an old plastic oil tank and he reckoned it would be just the job for them. He even went so far as to collect it for me and cut the front off it so they could get in and out. We cleaned it up with soap and water, put it in the new piggy plot and put concrete blocks on either side of it to stop it rolling. Then we filled it with straw. As it turns out, it worked pretty well – pigs aren't that fussy about their accommodation and as long as they have somewhere warm and dry and out of the prevailing wind, they are happy enough.

My neighbour has an old Land Rover which was ideal for the job of transporting them, so around mid-March we drove over to Wexford to collect them. I was actually quite nervous on the day. I'm always like that when I make a leap into the unknown: all the different (often conflicting) advice I've got from people is bouncing around in my mind along with all sorts of other dreadful scenarios: pigs escape and kill some children; pigs escape across a river and are a real nuisance for the rescue services. That sort of thing.

My first thoughts on seeing the piglets was that they were really cute. They were about three months old so they were basically Babe, even imitating that thing she would do with her head in the movie – you know, sort of cocking her head to one side. Very endearing. My second thought was that they were really LOUD. We had to put tags on their ears (more Department of Agriculture meddling) and you wouldn't believe the squeals they let out during that process. It's an earth-shattering noise that stays in your mind for hours afterwards. Pigs are basically drama queens – they are completely quiet animals when left to themselves (though I didn't know this yet), but if you make any move to catch them or carry them or mess with them in any way, they go berserk. At the time, as I stood there amidst the noise, I was very worried indeed. What if they make this much noise all the time?

Eventually we got them into the back of the jeep in which we had laid down some straw for them and they curled up together to get over their ordeal. There was more squealing when we got back to our house in Dunmore and we had to take out two of them. It was only about seventy metres from where we parked to their new home, but it was still an almighty struggle. I carried the female and my neighbour selected one of the males and they let out roars of indignation.

I put them into their new house, AKA the oil tank, and put a wooden pallet over the front opening to keep them in, temporarily, to get them used to the house. It was quite cold at that stage of the year and pigs don't seem to like the cold. They burrowed down into the straw, covering themselves up for warmth. After a few hours I let them out of the house and they explored their new environment. I'd say after fifteen minutes or so they had got out through the electric fence. I expected this because I had read on a website that it would take a few hours for them to get used to their boundary. Still and all, it was hairy enough trying to get them back in. Incidentally, the fence consisted of three strands of electric wire – the first about 15cm off the ground, and the other two in 15cm spacings above that. I've got loads of shocks off it and it's not pleasant, but it wouldn't kill you. When the pigs hit off it, they jumped and let out a little squeal, but mainly they steered clear of it altogether – eventually.

They got used to their boundary pretty quickly and got down to some serious rooting. And that's basically all they do all day long. Pigs root and eat soil because it gives them vitamins and minerals they need. They also eat small insects, grubs and worms that they find while rooting. I can't imagine what they would do all day if they were on concrete. The guy I got them from told me that he reckons pigs that are reared on concrete (which, I remind you, the vast majority of them are) actually go a bit crazy and who could blame 'em?

We fed them a mixture of organic pig nuts and rolled barley – the barley is left to steep in water in a big bin and it turns into a sort of a porridge. It's anybody's guess how much you should be feeding them because they never appear to be satisfied – you could feed them all day and they would keep eating – a fact not gone unnoticed by commercial pig farmers. I read somewhere that a good rule of thumb is to feed them enough to last twenty minutes, twice a day, so we tried to stick to this. They also got (and loved) scraps – leftover bread and pasta, sour milk or butter, fruit and vegetable skins and peelings – but no meat of any kind. Family and friends provided their scraps too on the promise of the odd chop. My Mum tells me that when she was young her mother used to keep a bucket for scraps outside the back door and a guy used to call around each week to collect them for his pigs. They used to call him Mickey Skins, so she resurrected the nickname for me.

My eldest sister, Kim, was the main scraps supplier, dropping off a bucket every other day. She has three kids, so there were lots of scraps, which was just as well because we wouldn't have produced enough ourselves. I kept telling her to give up if she found it a pain in the ass, but she swore that she enjoyed being involved and, besides, she would be putting it all in her brown bin otherwise and paying the council to take it away. (Incidentally, it's illegal to feed pigs scraps from the kitchen, so don't tell the Department.) The sow (I call her the sow, but I think the correct name for a female pig that has not yet had a litter is a 'gilt', but let's not get too technical) quickly established her dominance and when we put food in the trough she would puck at the boar with her snout. This meant that she got the lion's share of the food and, as a result, she put on weight much quicker than he did. By the time we came to kill them, she was much longer, taller and fatter.

Two pigs don't need as much space as you would think. I've talked to lots of people who are amazed at the fact that we were able to keep pigs in our garden. I think they were happy enough there too. Pigs love to break into a trot every now and then and they would run around the trees in the middle of the plot like they were doing laps. The trees also provided them with shade, which is important for pigs as they burn easily in the sun, and they also enjoyed rubbing up against the trunks for a good scratch.

They were also very playful. The phrase 'piggy back' apparently *does not* come from what our pigs spent a lot of time doing when they were younger (putting their front legs up on the other's back and riding around for a while) – but it should do. I thought, at first, that maybe it was a harbinger of some imminent sexual shenanigans, but then I noticed that they seemed to be taking turns. It seems to me the sole reason for this strange activity was that it was FUN. I also suspect that pigs have a streak of what I can only call 'divilment' – they liked tipping their water trough over just after I'd refilled it and seemed to enjoy the sight of me going over and back across the garden with buckets of water. Like some of our European neighbours, our pigs took regular siestas. On cold days they stayed indoors covered in straw, and when it was nice and warm they would lie out sunbathing.

They had the little area which we cordoned off for them completely stripped of grass, weeds and everything else within a month or so. After that, I think they started to get a little bored and they would eye the lush lawn on the other side of the fence covetously. They would also test the fence every now and then with their snouts to see if it was still on. One afternoon I was looking out the window of my office and noticed the two pigs out grazing on the lawn. The stupid batteries had gone dead on the fence and, cute hoors that they are, they were out straight away. That time they were only out for a few minutes, but about a week later I was away for a few hours and when I came back they were not only out, they were over at the other side of the garden and had done some serious damage to the lawn, rooting to their heart's content. This time around, the fence had shorted out because they had rooted soil up on top of the bottom wire. We were feeling pretty uncomfortable about these attempts at liberation because if we went away for a whole day and they got out, we would have no garden left.

So, it was back to the co-op, practically embarrassed to be looking for *more* fencing posts. I also got some sheep wire, about three feet high and put that up around the outside of the posts. The electric fence was still there, but the sheep wire acted as a more robust, second line of defence and they never managed to breach it, and we were much happier for it. They were definitely smart enough to test the electric wire on a continuous basis to see if it was still active ... organised pigs, now that is scary.

Pigs are easier to get attached to than hens. After a few days acclimatising they started to get quite friendly. They seemed happy to see you any time you came near them. Of course, this is probably an illusion and they were just happy to see you because you're the guy who brings them food. Dogs are much the same, but we just put human emotions like fealty and loyalty on them to make us happy. Whenever we went out the back door and closed it behind us, we could see the two pigs pricking up their ears down at the corner of the garden, listening, no doubt, to see was their food on the way. I used to find that kind of charming.

After a couple of weeks they would happily stand and let us give them a rub behind the ears or a scratch on the belly. You could get into the pen with them, but you didn't hang around in there or anything, especially if they were hungry. They are intelligent creatures, but they aren't great at working out what's food and what's not. I'm not even sure they care. They bite at everything. A welly. A leg. The bucket. And pigs are strong creatures – I've seen one of them get their snout in under their house and flip it over with one flick of the neck.

They were unobtrusive inhabitants of our garden most of the time. Unlike the daily dramas with the hens, pigs are pretty self-sufficient. They did a bit of squealing in the morning and evening if they were really hungry or if they saw you coming with the bucket. But it was nothing like the squealing I'd heard the day we picked them up. They are fairly noisy eaters, slurping and slopping, but I know a few people who have worse table manners. They give out a highly amusing appreciative grunt when they are eating something they really like — it's a low sort of a grunt that for some reason always reminded me of Winston Churchill. I don't know whether Churchill actually grunted at all, but I always felt that he must – probably, I imagined, the impact of all those cigars on his sinuses.

There's one myth that I can dispel here and now. Pigs are not particularly dirty animals. Ours kept their house completely clean and tended to shit down the other end of the pen all the time. I don't particularly want to have a big talk about the composition of pigshit, but let me just say this: it looks a lot like dogshit. I kind of assumed it would be messy, runny stuff and that they would roll around it. You know like the saying, 'happy as a pig in shit'? But it's not like that at all. If there was a lot of shit around and it rained, there was a bit of a stench down there, all right, so I would go in every other day with a trowel and bucket and take it all out – not the greatest job in the world, but it's dynamite on the compost heap.

Unfortunately, the weather was absolutely abysmal and June, in particular, was the wettest on record. The ground they were on basically turned to sludge with the amount of rain that fell. This was a pity because I don't think they particularly liked the mud and spent a lot of time in their house. It also meant that things got quite smelly down there for a while.

We didn't really get into naming them or allowing ourselves to get too attached to them. This was important because they were only to spend about five months with us and we didn't want to lose focus: these were animals we were rearing to eat. You don't want to be getting all rainy-faced about it. My two nieces came up with ridiculously cute names, Charlotte and Wilbur, when we got them first, but these names never really stuck, thank God. I wanted to call them Sausage and Rasher, but that just seemed to be rubbing their noses in it. So really we didn't call them anything at all.

We had them about three months when I started getting worried about possible sexual shenanigans. Our plan had been to kill them in July or August, but by May they didn't seem to be fattening all that well and I figured it was only a matter of time before they started 'at each other'. We didn't want to be under pressure to bring them to the slaughter prematurely, especially since they still had hams 'like a duck's arse'. So, reluctantly, we got a vet out to sort out the male. A more unpleasant experience you could not have – for us and him.

You don't appreciate how strong an animal is until you're trying to catch it to cut its balls off. There was more earth-shattering squealing (the type that stays in your head for days) and he kicked like an irate mule. Eventually Mrs Kelly and I got him pinned down, her at one end, me at the other and while he squealed indignantly the vet gave him a local anaesthetic, made a quick incision in each scrotum and then removed the testicles with a quick yank. It was completely and utterly gross to watch, but it was over in seconds and he seemed none the worse for the experience. Within minutes he was chomping away on some food we put out for him. He was walking a bit funny for a day or so, but then, wouldn't you? After that he was fine. I was really pleased we had done it – horrible and all as it was – it meant we could hold onto them until late summer if needs be without worrying about a brother and sister mating and producing a litter of mutant piglets. Next year we will definitely get two sows.

In the end, they went to the slaughter in July. They put a right spurt of growth on in late May and June and my father-in-law came to inspect and professed himself happy with their progress. The whole slaughter thing happened very quickly, which was good because it meant we didn't have time to dwell on it too much. I rang one of those very rare, small, local abattoirs discussed earlier, just to get some information and the woman I spoke to told me that they were killing pigs the following week and if we didn't get them done then, she wouldn't be killing again for a month. We reckoned they would be too fat by then, so we decided to go for it. The owner of the abattoir really made me feel at ease – she patiently answered any questions I had and never once said: Jesus, you really are clueless.

I asked her could I go along to see it being done and she said absolutely no problem. I don't know why I thought this was a good idea. I guess I just felt that the reason I had got involved in rearing pigs was because I wanted to reconnect with the realities that lie behind eating meat – the reality of what's involved in rearing them *and* the reality of what's involved in killing them. I reckoned that after everything we went through together, it would be pretty cowardly of me to drop them off at the abattoir and just drive away and not see what happened to them before they were diced and sliced.

She suggested to me that we should try and minimise the

stress for them on the day of slaughter. She asked how we were transporting them and I told her that my father-in-law had kindly agreed to provide a trailer. 'You should get him to drop it down a few days beforehand if possible,' she said. 'You could feed them in the trailer or let them sleep in it for a few days. That way you won't have any hassle on the day.' So we did just that and they were delighted because it was roomier than the old oil tank. They were so happy, in fact, they barely stirred all weekend. It was like they were having a luxury weekend at the Hilton for their last days on earth. It did make things easier on the Tuesday morning when I was taking them to the abattoir. I put their food in the trailer and, while they were eating, closed up the back and that was it – off we went. We got to the abattoir at about lunchtime and unloaded them into a pen prepared for them. There was one other pair of pigs there at that stage who were to be slaughtered that evening too – I was worried because they looked about twice the size of our ones.

They were scheduled to be killed at 7.00pm so I went off for a few hours. When I came back they were sleeping away in their pen, completely oblivious to their fate – exactly the way you want them. It's important to mention that the next few paragraphs are going to be pretty gruesome, but I am not trying to be gratuitously disgusting or anything. I wanted to be there to see them killed so I would know exactly what happened to them – but you might not necessarily want to read about it, so if you don't, you should just skip the next two pages.

The butcher arrived with his tools. The pigs were taken about twenty yards from their pen to a holding area where they were separated. The sow was first up. I patted her on the head, I suppose because I felt I should mark the moment somehow. Then I thought to myself, that's ridiculous and actually kind of condescending. But apart from that little blip, I didn't feel any emotion, really. The butcher came up beside her with a device that looked like a garden shears – this is a stunning tongs and it delivers a shock of about 200v to the pig's neck. She went completely rigid and when he pulled it away, she just fell over to the side, much like our traumatised hen after the fox attack. She wasn't dead, he told me, just unconscious. There wasn't so much as a squeal out of her, which was a relief (I remembered the noise the day we did the job on the boar and I presumed that the slaughter would be at least as bad). She was then hoisted up by her back legs and he stuck her in the jugular with a knife. Simple as that. The blood drained from her and she wriggled a little, but bear in mind she was unconscious at this point so she felt nothing. And that was that. I had expected my heart to be pounding in my chest as it was when I killed the chicken, but actually I watched it all quite impassively. I don't know why.

Once the blood had drained out and the pig was dead, she was put in a large machine, which is filled with boiling water and spins the animal around. This softens the skin and removes the hair. When the butcher opened it up again after a few minutes, I was amazed to see my pig looking like any other pig – pink as opposed to ginger. The pig was hoisted up again, washed and then he used what looked like a mini-flame thrower to burn off any bacteria that might be lurking on the skin. Next up, he slit the pig from top to bottom in one impressive cutting action, removed the stomach, liver and intestines – this was the only point in the whole process where I felt a little queasy, especially because he dumped them in a wheelbarrow right beside me. Then he used a saw to cut through the breast bone and removed the heart and lungs. Finally, he used an electric saw to cut the animal in half from top to bottom. And that was it. The carcass went into cold storage in two halves to await butchering. The process was repeated on our boar and six other pigs that night. It was interesting to note how much smaller the boar was than the sow – all her bullying at the trough had paid off. You could also see bruises on his neck where she had been pucking at him during feeding. The auld bitch. Our pigs were small compared to some of the others, but I was pleased to note they weren't the smallest and the owner of the abattoir said they were a good size for Tamworths.

There are a couple of things I thought about that evening which are worth recounting, I think. First of all, pretty much ALL I could think of as I watched him take out the entrails and dump them in the wheelbarrow was how connected the food we eat is with the food the animals eat. I mean, the scraps that I fed them that morning were probably still there in the intestines. I decided there and then that I would never eat an animal again if I didn't know what it had been fed. (That promise didn't last very long because when I got home that night we had a very tasty steak which had been bought in a supermarket! Still, I *am* trying to implement my vow.) The other thought that I had was about the butcher himself and what amazing skill he had. It would probably take me about three days to gut an animal that size but he had it done in seconds. All these skills will be lost forever if the government continues to close down the small, local abattoirs.

Before I left the abattoir I sat down and discussed with the owner what we wanted done with the meat. A few weeks previously I had thought about getting the carcass delivered and trying to butcher it myself (with the help of a friend's dad, who is a retired butcher). I had dreams of making my own pudding and sausages, curing my own bacon – those sorts of things. Thankfully, common sense prevailed. We might do it in the future when we are a bit more clued-in. In any case, you are not allowed to take offal or blood away with you from an abattoir anymore, so you can't make your own pudding. For €80 per animal the abattoir took all that hassle off us – they did all the butchering, cut the meat as instructed, cured the bacon, made the sausages up for us – all that stuff. And because they are a small, family-run butcher's, we were happy for them to do that. We decided to dedicate one pig to pork and the other to bacon – the basic difference being that pork is just pure pig meat while bacon is cured.

Two weeks later I collected five large bags of meat. As I said I'm a big fan of sausages and rashers and was pleased to see we had about 10kg of each, enough to see us through four or five months. There were joints of pork and ham, pork chops, ribs, pork steak and diced pork. We even got back the trotters and heads, although both are still in the freezer and I am not sure I have it in me to eat them. We have an old, six-drawer freezer in the garage which we completely filled with meat as well as the freezer attached to our kitchen fridge. We even had to give some of the overflow to the mother-in-law to put in her freezer. I haven't worked out the value of the meat we got, but at a guess I'd say there was about €500 worth there; 10kg of premium sausages alone would cost you the guts (excuse the pun) of €100. I reckon it cost us about €350 in total to keep the pigs (including buying them, then feeding and slaughtering) so I suppose we made a small profit. More importantly, though, we were very happy with the taste of our meat – and the sausages from the off-cuts are great, which for me was the greatest test of all. I'm not a big fan of pork chops, but these ones taste superb – although they are a little fatty and have to be trimmed before cooking. The rashers have a nice bit of fat on them too and taste great – the butcher cut them nice and thick, so they don't shrivel to nothing when cooked. So far, we can discern no difference between the sow and boar meat. We don't feel strange about eating the meat although Mrs Kelly doesn't like me to talk about the pigs when we are actually eating them. I'd say something really stupid while tucking into a chop, like, 'Ah sure, the old pigs were great, weren't they?' and she would go mad.

We have given a lot of meat away and the feedback is very positive – we gave a collar of bacon to a friend of ours and he said that if that's what bacon is supposed to taste like, he's going to get some pigs for himself. Another friend said that the sausages taste like the way she remembers them from when her granny used to cook them up for her when she was young. Anyone we have given pork chops to, seems to be particularly impressed – I suppose the pork chops that we buy in supermarkets these days are fairly bland so ours taste great by comparison.

I didn't enjoy anything about the slaughter process. It was unpleasant to watch, but I'm glad I was there. I think I respect the meat that we have more because I knew the animals in question and because I was there to see them killed. That probably sounds strange, but it's true.

All things considered, I cannot imagine why we would not want to repeat the whole process again. It's great not having to buy any pork at the meat counter in the supermarket – there's a definite reduction in our weekly meat bill (I would imagine between sausages, rashers and ham I used to spend at least €15 a week on pig meat). And it's not just that we don't have to buy pork, but also we're not inclined to buy much other meat. The logical next step would be to get into breeding from a sow so that you wouldn't need to buy the pigs at all. I would love to do this, but I don't think we have enough space and I think we would have to get some more permanent housing if we were to keep them over winter. The oil tank was great and all, but in the winter I think they would need a proper house. In addition, while the little patch down at the end of the garden worked well in spring and summer, it would be a complete mess over winter.

It would be perfectly feasible to carry on keeping a pair of weaners each spring, fattening them up until late summer or early autumn, and filling the freezer each year. I would like to try a different breed next time around – the Tamworths were great but I think the meat was a little too fatty for my liking. Mrs Kelly and I were down in Ballymaloe recently, walking around the cookery school gardens, and in a field out the back I found a litter of saddle-back pigs with an enormous sow – they looked like an interesting breed and I think we might try them next. Because our pigs had long since been slaughtered, I found myself standing there looking at them for longer than is healthy. I guess I missed our pigs.

During the winter months I get the odd pang when I walk out the back door, look down the end of the garden and see an empty plot down where the piggies used to be. It's certainly not regret or remorse about killing them, and it's somewhat of a relief not having to feed them morning and night – but I did enjoy having them around.

I wasn't the only one who missed them. One of my little nephews visited us recently – he had been down when the pigs were alive and had got quite attached to them. The first thing he did when he came this time was run down to the end of the garden to see the pigs. He was absolutely disgusted with the news that they had been killed, especially when he discovered *why* they were killed. It was the first time that he made the connection between the food on his plate and the animals that he loved to play with. For a few weeks he wouldn't eat any meat at all. My sister told me that one day he took out a cookbook and went through each page with her, pointing to the picture of a meal and asking, 'Does *that* come from an animal?' Yes, she would say, and he would move on to the next recipe and ask the same thing. Around that time our springer spaniel Ozzie died and the poor little boy asked his mother were we going to eat him too. The first time I saw him after that he wouldn't talk to me for about twenty minutes, hiding behind his dad's legs so he wouldn't have to look at me (or perhaps afraid that I would grab him and take a big bite out of his leg). He got over it eventually, of course, as all kids do, but it was an interesting study in how difficult we find the whole area of killing animals for food.

HARVEST

There's a fascinating concept that I've come across recently called biophilia – the term was coined by a philosopher called Erich Fromm to describe mankind's love of life and living systems. I don't profess to know all there is to know about biophilia but here's my understanding of it. A 'philia' is the very opposite of a 'phobia' – it is an attraction towards some-thing in our surroundings. The concept of biophilia suggests that we subconsciously seek out interactions with other plant and animal life forms and the reason we do so is that we are hard-wired that way. In other words, our brains have evolved to need interaction with things that are alive and vital.

Apparently this is all part of Mother Nature's cunning plan to ensure that life thrives on the earth by pre-programming mankind with a basic desire to like other life-forms. This makes sense when you think about it – I mean, if we were pre-programmed to absolutely loathe plants and animals, we probably wouldn't rest until every last one of them was destroyed (though some would argue we're well on the way). This, then, is the reason we love nature – the reason we like having potted plants and doomed bonsai trees in our homes, the reason we keep cats, budgies, parrots, tarantulas and snakes as pets, the reasons we smile when we see a cute puppy on the telly playing with a roll of toilet paper. We can't help it – it's a product of our biological evolution.

The fact that our brains are wired to derive joy from interacting with plants and animals perhaps helps explain why many people feel so utterly alienated in modern society. How could we not? We have laid concrete and tarmacadam over vast swathes of the earth to make way for cars, homes and businesses. We can live out our daily lives, if we want, without ever coming into contact with a living thing (other human-beings aside). Most of us will go through life without ever sticking our hands in the clay to pull out a vegetable to eat. We have no notion of what it's like to care for and look after a beast that will one day become a meal for us. Those basic interactions, which we have evolved to need, are absent from the vast majority of people's lives, replaced by an industrial-ised food-chain underpinned by unscrupulous producers and driven by the desire for profits.

To my mind, biophilia helps to explain why Mrs Kelly and I have got such immense satisfaction from the tiny steps we have taken towards self-sufficiency since we moved here. It also probably helps to answer the question I asked earlier about the Downsizer's Dilemma: after I've cashed in that inheritance cheque from my rich aunt will I still be bothered with growing vegetables or rearing animals? Of course I will, because this lifestyle is about far more than saving money. When I look back on all the changes we've made over the past three years I think they were all driven by a desire to have a simpler life, something closer to the way things used to be, to the way things were for our parents or for our parents' parents. Perhaps that's a slightly naive pursuit in modern society, but still, you can try. We were watching the nine o'clock news over with the in-laws recently and we were giving out about the way the country has gone – how could you not grow weary of the relentless bad news: the murders, gang-land violence, drunkenness, suicide? I was asking my father-in-law whether, if he had the choice, he'd prefer to be living in present-day Ireland or Ireland of the 1950s and he said, without hesitating, the fifties. I said to him: 'But doesn't everyone say that? I mean, if you asked your parents the same question back in the fifties, would they have preferred the early 1900s?' He said they wouldn't, that they had hardship when they were growing up that we can't even comprehend today and they probably would have preferred the fifties too.

Was life really all that rosy back in Ireland in the 1950s and 1960s? Let's not kid ourselves by saying that everything was great back then, it wasn't. But there is a sense that perhaps it represented a time when we were perched between two extremes. There wasn't that grinding poverty that existed earlier in the century, but at the same time there wasn't the same complexity that we have now and there weren't the same obvious levels of naked greed and aggressive materialism. There was common decency, in spades. People knew each other and helped each other out. Community was everything. My father-in-law talks about how farming back then was a community activity, not a solitary one. When there was a big job to be done on the farm, neighbours came around to help. People pitched in. Today we have fooled ourselves into believing that we don't need anyone but ourselves to get by – that once we have our jobs and our homes and our surround-sound tellies, we don't need anyone or anything else.

There's a perennial argument in the media about which is better – modern, confident, vibrant Celtic Tiger Ireland or simple, backward, Church-ridden, poverty-stricken Ireland of old. I don't think it's a black and white issue. Certainly there are things about modern Ireland that are far better than they used to be. It's great that people have good jobs, can afford homes and that our young people don't have to emigrate. But there were also things about the old Ireland which were far better than the way things are now. I think there is a very large group of people in this country who feel that at some point things just went too far. We didn't know when to say 'enough'. They are people who haven't found happiness in the things modern society tells us we must have – bigger houses, flashier cars, more success, more money. I come across these people all the time: people I interview for 'A New Life' in the Tuesday health supplement, for example, who have given up their jobs and turned a hobby into a career or moved back to live in the countryside; people who have given up affluence in exchange for happiness; people wandering around farmers' markets or growing some vegetables, trying to get back in touch with the simplicity of country life, or getting back to cooking and eschewing convenience foods. People who believe that less is more and that the very foundation of our society – that *more is more* – might not be the road to lifelong happiness.

Ironically, it seems that the 'Giving Up' column was itself an attempt to recreate the 1950s right here in the noughties – trying to turn the clock back and imagine a world without cars, mobile phones, TV, skinny lattes, deodorants, dishwashers and razor blades with silly names. Perhaps it's not surprising in that context, that the 'Giving Up' experiment that I was most proud of was also the one that most accurately recreated what life in old Ireland was like. In the winter last year we gave up electricity for a week. Poor Mrs Kelly had to endure the full brunt of this experiment, not just watch me making a fool of myself. The week that the feature appeared in the magazine, the *Irish Times* gardening guru Jane Powers was guest editor and it was a special Green issue – so my experiment seemed to tie in nicely. It was an odd experience – in some ways it felt like taking a very big step back in time, though given concerns about peak-oil and the fact that energy costs are starting to become such an issue, it was also strangely cutting-edge.

Giving up electricity was really a whole series of experiments rolled into one – giving up light, heat, showers, the oven, fridge, kettle, toaster, TV, radio, house alarm, dishwasher, microwave, vacuum cleaner and all semblance of modern life. On the Monday morning I turned off the mains at the fuse board, just so there could be no cheating and lit the solid-fuel stove in the sitting room, since it was winter and quite cold in the house. With the flick of the mains switch, an eerie silence descended – no background noise of TV or radio, no gentle hum of the fridge or computer. We did the experiment during the winter because we felt it would be just too easy in the summer. We also tried to keep our normal routine going. For example, we reckoned it would be cheating to start doing things at night that we normally wouldn't do then just to get away from the cold, dark house, eg visiting our non-electrically challenged neighbours for tea, going to the warm, bright cinema every night, heading to the leisure centre for a hot shower. The idea was to try and maintain our modern lifestyle without the juice that usually powers it. In that spirit, on the first morning I did some hand-washing in the kitchen sink using water heated on the stove. This was a complete pain in the ass – it took about an hour to warm the water in the first instance and then another hour to do the washing. The experience of hand-washing Mrs Kelly's 'smalls' will live long in the memory. God bless washing machines.

Each evening when darkness descended, I lit about twenty candles and an old oil lantern I found in the garage, but even with all of them lighting it still seemed really dark. One of the great things about the experiment was that it highlighted areas of 'peak-oil' vulnerability (if you believe in such a thing, and not everyone does). Take our cooker, for example. When we moved here first we had a big debate about the pros and cons of buying a range. On the plus side, we felt it would be appropriate for an old cottage and would feel very homely. But I didn't want to buy an oil-fired one, and this meant we would have to get a solid-fuel version. Now, although there's something very attractive and caveman-esque about the idea of lighting a fire to cook your food, even I have to agree that the practicalities of same would be fairly grim. In the end we opted for an electric range with a gas hob, which sort of looks like an Aga if you squint your eyes a bit or fill up on beer. But oh, how we would have loved the comforting warmth of an Aga that week. Then, when we bought the stove for our sitting room shortly after we moved in, we decided against getting the model that would fire the radiators in the house – mainly because it throws less heat out into the room, and since our sitting room is big we needed the stove to heat the room, not the rest of the house. Anyway, that decision too seemed pretty short-sighted that week. We kept the stove lit 24/7 for the whole week so the sitting room was nice and warm, but without electricity to fire the oil burner, the rads in the rest of the house were cold. It really does make you wonder what would happen if oil suddenly became prohibitively expensive? How many of us would be able to heat our homes?

Anyway, if the heat (or lack thereof) was an issue, the lack of light was an even bigger one. You discover very quickly how much of a struggle even the most basic activity is when your world is lit by candles. Most of us love the idea of candles in a kind of romantic, get-yourself-in-the-mood-for-luv kind of sense – but they are really, really shit at fulfilling the task they were invented for, namely lighting up a room. Try, for example, to cook up a pot of spaghetti Bolognese in the semi-darkness and you'll see what I mean. In an attempt to see what was going on in the saucepan, I held a candle over it and some wax dripped in. Was it because it's so hard to prepare a meal in the dark that people used to eat dinner in the middle of the day? We pondered this question while eating our waxy spag bol and then retired quickly to the sitting room, the only place in the house that had any heat. We pulled two chairs up to the stove and sat there talking. I tried reading the paper, but gave up after ten minutes, half-blind from squinting in the dark. We played cards for a while. Talked a little more. 'It must be time for bed,' I said. 'It's only half-seven,' replied Mrs Kelly.

The darkness was oppressive. We had to carry the lantern with us wherever we went (to the toilet, for example) and even then it only threw a little bit of light immediately out in front of you. Someone suggested that I try getting my hands on a Tilley lamp, which, apparently, was widely used before electrification and, I think, burned paraffin, but there didn't seem much point since we only had a week, so we ploughed on with our candles. When you think about it (and I had lots of time to think about things that week), it's pretty impressive how electricity has helped us make an irrelevance of the seasons. We carry on with our hectic lives in the depths of winter using artificial light and heat, but without electricity the darkness and cold of the winter months must have forced people to slow down at that time of year. Perhaps that semi-hibernation was how nature intended it?

You also get a sense of just how short the winter days must have been back then – eight hours of light followed by sixteen long, dreary hours of darkness – and how excited people must have been in the spring when longer days arrived. We talk about SAD (seasonal adjustment disorder) these days, but imagine how much worse it was before electricity. Or maybe it exists as a 'disorder' only because we try to pretend that seasons don't actually exist?

We also found the darkness that week to be very eerie, and, dare I say it, almost spooky. There were shadows everywhere and those corners of the house that had no candle burning in them seemed almost off-limits. You can just imagine the effect that ghost stories must have had in the old days, told around a fire in the depths of winter. You could understand how people were superstitious and more aware of the supernatural than we are in our lovely, bright winter homes. Imagine hearing a story, say, about a banshee in a candlelit room with no chance of any light entering your world until the next day? Sure, you'd be shitting yourself.

We made a shrewd, strategic switch of bedroom early in the week, which reminded me a lot of when I was a kid and we used to move into our 'winter sitting room'. That makes my family sound extremely opulent, which we were not, but we did have two sitting rooms and for some reason when the winter months came we would decamp for the season to the 'good' room – I used to love that because it always meant that Christmas was on the way. Anyway, there's a little bedroom to the front of our house which is directly over the sitting room and it's always lovely and toasty at night when the stove has been on. Our bedroom, which is at the back of the house, is cold at the best of times – it was freeze-your-nuts-off cold that week. In bed at night I tried reading a book by candlelight but was consistently conscious of it being a serious fire hazard and eventually the candle extinguished anyway and the room would be suddenly plunged into darkness. For the first few nights I woke up in the middle of the night and felt my way (nervously it has to be said) downstairs to put fuel on the stove so that the house would be a little warmer in the morning. It felt like the kind of thing an Irish mammy might have done fifty years ago. Out the back door I saw the most magnificent array of stars in the sky – they are always pretty impressive on clear nights but with the house blacked out they were incandescent.

In the mornings when the alarm went off (battery-powered) at seven o'clock our bedroom would be pitch black and Baltic cold. Fumbling for matches in the dark, you would long for the simplicity of flicking a switch and throwing light on your world. In the absence of electricity the day awakens more slowly. Because of my selfless Irish mammy refuelling job, the pot of water on top of the stove would still be warm and we used it for washing ourselves. Washing was done at the sink with the pot of hot water and a facecloth. We have a framed chalk drawing in our bathroom of a little kid washing his hands with a facecloth, which I always considered cute until I had to do it myself. Still, I was pleasantly surprised to find I felt quite clean afterwards. Later in the week we had a bath – it took about two hours to get enough hot water from the stove and gas hob, but it was worth it – when you haven't had a shower in five days, you'll go through any amount of preparatory work to immerse yourself in hot water. Because our bath is not really big enough for two, and because it was too cold to be getting amorous, we bathed alone, taking turns to get in. We drew lots on who got to get in first, which I lost and therefore had to go second. Bummer. After sharing Mrs Kelly's used bath-water and washing her smalls in the sink by hand, I think our relationship has moved permanently onto an entirely new level.

Mealtimes weren't really all that different – as I've said, we still had the gas for cooking and we had the stove for doing stews or anything that didn't mind (really) slow cooking. But we missed the oven – I have quite a thing for baking bread each day, which I obviously couldn't do, and again I was thinking about how nice it would be to have an Aga.

I found I appreciated the daylight hours more than normal and tried to cram in as many things as possible, especially with regard to meal preparation. I would prepare a stew during the day and put it on top of the stove in the sitting room, where it would take about six hours to cook and the heady aroma would fill the house. The fridge was out of action but that wasn't as big an issue as you would think – the vast majority of things in our big, modern fridges don't actually need constant chilling: beer, wine, chutneys, marmalade, fruit, vegetables and so on. How ridiculous is it to put chutney, which is all about preserving food anyway, using vinegar, into a fridge? We put butter, cheese, milk and any other perishables out in the porch to chill, where it was about -20°C anyway. We bought meat as we needed it, which meant more trips to the shop than normal – again an accurate representation of what life was like in the 1950s. (We didn't have our freezerful of pork at this stage.)

The week dragged on interminably. By day I worked in my office, which is separate to the house and was still powered up – well, a man's gotta work. It was a little island of electrification, warm and bright, with things beeping, buzzing and whirring. There was an inevitable, depressing relentlessness about the arrival of the darkness each day and my return to the cold house. I couldn't get over just how ill-equipped we were to deal with the long hours of darkness. There's no doubt in my mind that television has made it very difficult for us to entertain ourselves in its absence. Certainly at the start of the week we were really struggling. But things improved. I tried to do a jigsaw one evening, but it was just too dark. We played cards, Scrabble and Monopoly, and I even tried to teach Mrs Kelly how to play chess another night – but I'm useless at chess and didn't know all the rules. We played charades, which is fun, if a little pointless with only two people. I played the guitar quite a bit and we both played a little piano – well the piano is regular size, oh you know what I mean. Another night I spent forty-five minutes down on the ground in the kitchen wrestling with the dogs, just for something to do. A lot of the time we would just sit staring into the flames in the stove in silence. After a while that becomes quite relaxing, but at the start of the week the silence was deafening. The stove was like a celestial object sitting in the corner radiating light, heat and goodness. It was our best friend, the star performer. We sat close to it and each other.

There is not a hope in hell that I would ever have considered repeating the whole exercise, were it not for one thing.

For our final night we invited friends and family around for storytelling and we had such an amazing night that we vowed at the time to do it again (NB, the storytelling night, not the whole week *sans* electricity). I was particularly keen to invite people who might remember what life was like before widespread electrification, to get people talking about how much life has changed since. We had a truly magical evening; it was far from a fancy dinner party, more a bowl of stew on your lap in front of the fire sort of thing. There was plenty of drink to warm the body, stories were told about ye olde Ireland, songs were sung. The mother-in-law even got up and did some Irish dancing. Anyone can have such a night, of course, but there was something about the darkness in the room and the context of the whole thing that made it special.

Particularly striking were the consistent references to how much more conscious people used to be about consuming electricity. When it arrived first in Ireland it was seen as an expensive luxury rather than a utility – and people were careful with it. They turned off lights when not in use, switched off the mains when going to bed. Environmental concerns and increasing energy costs may force us to revisit that frugality. Someone told a story about how there used to be ice on the *inside* of windows in their house when the winter turned really cold – now that's hardship. The general consensus at the end of the night was that it should become an annual event, though I'm sceptical as to whether it would be as good, second time round. Plus, the washing-up afterwards was a complete nightmare.

The following morning we luxuriated in a long, hot shower. Candles were put away. Lights went on. Radiators heated up. The electricity metre whirred into action again. I went out to the office and tried to write about what we'd learned. I was tempted to go down the route of saying that giving up electricity for a week is perhaps a good way to see what life might be like in the future if you are to believe the soothsayers who say that peak oil will make electricity so expensive that ordinary Joes like us won't be able to afford it. I'm not inclined to think that things will get that bad, mainly because if you were to worry about things like that you'd drive yourself mad.

So, I thought it would be better to focus on the positives that we took from the week and they were this: we certainly re-learned to enjoy just being *with* each other, talking, entertaining ourselves rather than relying on the goggle box. Being still. There must be longterm value in that. Sitting by the fire at night with nothing to do but talk was kind of boring at first, but grew less so as the week went on – that leads me to think that being sociable and being able to entertain ourselves is a skill that we have all but forgotten, but, thankfully, it is something we can relearn. It's like a flabby muscle that can be coaxed back out of retirement by regular exercise.

The other positive was the storytelling night which was an absolute highlight because it was a night of family, friends, food, banter, stories and song – a wonderful, close, ancient, simple get-together that warmed the heart and the soul. It comes back to what I was saying about community – it's unlikely we would have been able to have a night like that when we lived in Gorey and knew nobody within a thirty-mile radius. In fact, the most memorable times we have had since we moved down here have been exactly those types of get-togethers – events where we are reaching out to our community and being hospitable. They remind us of how lucky we are to have family and friends living nearby. Shortly after the pigs met their maker we had a harvest feast of sorts – I love the idea of reclaiming the harvest as a time of celebration. These days, late August and early September can be a pretty depressing time because we're back at work after summer holidays and kids are back at school. In any case, the harvest is not something worth celebrating in an environment where any food you like is available no matter what time of the year it is. But traditionally harvest was a time to celebrate the fat of the land with family, friends and neighbours. Growing some of your own vegetables allows you to re-possess the harvest celebratory feeling because in the late summer you genuinely have a harvest to deal with – there are more tomatoes than you can eat, big fat courgettes crying out to be picked, fruit trees laden down, peas coming out your ears. Along with hours spent in the kitchen trying to process them in some way to use them up or preserve them – chopping, pickling, bottling, freezing etc – it's also a nice idea to invite people around and have a big nosh-up.

I had in mind for a long time that, once we had the meat back from the abattoir, it would be an appropriate juncture for some sort of a celebratory feast – a sort of thank you to the pigs for their efforts and a celebration of the fact that for the first time we could churn out an impressive feast, meat and all, using only produce that we had reared or grown ourselves. The menu for our feast looked like this: pork chops, marinated in our own sage, our own garlic (stop me if the 'our own' stuff is getting a bit tedious, but it gives me a great thrill to repeat it) and some olive oil, which clearly wasn't our own. These were cooked on the barbecue along with some sausages (our own). We also had a cooked ham from Charlotte (or was it Wilbur, I can't remember), which was boiled and then glazed in the oven with brown sugar and served with crusty, homemade white bread. We had two different salads – a cucumber and dill salad (a handy way of using up a glut of dill) and a salad of basil, lettuce leaves, baby chard leaves and tomatoes. We cooked some veggie skewers on the barbecue with courgette, cherry tomatoes, red onion and green pepper. There were also freshly dug spuds boiled with fresh mint. All our own. For dessert there was a plum tart, which used up lots of plums, with ice-cream – the ice cream wasn't ours, but it came from Kilkenny so it hadn't far to come. The masses came, and went away satiated. It would finish the paragraph perfectly if I could say that this wonderful culinary extravaganza was washed down with lashings of home-brew, but unfortunately my home-brew was a complete disaster. I should explain.

A few months ago, I interviewed Will Sutherland, who now runs the late John Seymour's School of Self Sufficiency in New Ross, County Wexford. Over lunch at his house, Will brought out a particularly fine ale and I had a small glass – it was absolutely gorgeous – a very soft, velvety brew with only a slight fizz. I don't even particularly like ale, but I was blown away by this and I asked him where it came from. To my amazement he said it was his own and as we talked about it, I got the familiar feeling of excitement that I always get when I feel a hare-brained project welling up inside me. He told me that he could never understand why more people don't make home-brew – that with the modern brewing kits it's the easiest thing imaginable, phenomenally cheap and produces beer that's easily as good as, or better than, what you buy in the shops. Will, you had me at hello.

He told me about a guy in Waterford who sells the kits so that very afternoon I called him and went to buy all the gear. You get a very large fermentation bin which holds about forty pints; a beer-making kit which consists of a hopped malt extract and yeast; and a decanting siphon tube, which you use for bottling the beer. The whole lot came to about €30 – for future batches all I will need is the beer-making kit which costs €17. Forty pints for €17 – you do the maths. In anticipation of our harvest feast I spent a pleasant few hours making the beer one night when I had nothing else on. You pour the contents of the kit into the fermentation bin and add six pints of boiling water. Then you add a kilo of sugar, give it a stir and fill the bin with cold water. When it has come up to room temperature, you add the yeast and put on the lid. You then leave it in a warm place to ferment for about a week.

When the fermentation is complete you siphon the beer

off into bottles. Will told me that two-litre 7-Up bottles are the right job for this – firstly, plastic is ideal because you can squeeze the bottle to check for too much fizz (the precursor to a beer explosion and a messy divorce), which you can't do with a glass bottle. Secondly, the colour of the 7-Up bottles (green) lets in just the right amount of light apparently. Anyway, we don't drink 7-Up in our house so I went out and bought about twenty bottles and poured the contents down the sink – you can only imagine the row that ensued when Mrs Kelly caught me doing this. What can I say? Sometimes my impatience to get started on a project gets the better of me. Siphoning off the beer into the bottles is actually pretty cool – I remembered how to do it because when I was young I recall my dad siphoning off petrol from the car for use in the lawn mower – the trick then, as now, was to keep the bottle below the level of the bin to allow gravity to do its thing and then to suck a little on the siphon hose to get the liquid moving. Suck too hard and you have a lot of beer in your mouth (which, let's be honest, is a lot nicer than a mouthful of petrol). Before closing the bottles, you add a little sugar to get the secondary fermentation going, and then comes the worst bit: you have to leave the bottles stand until the beer is clear. This takes two to three weeks.

I got tremendous satisfaction out of the brewing process itself but, unfortunately, the beer tasted completely disgusting. One of our neighbours said it tasted just like normal beer, which was very kind to me but very unkind to the brewing geniuses of the commercial world – everyone else said it tasted 'just like shite'. Anyway, I'll persevere with the home-brewing because, again, there's something profoundly, hic, comforting, hic, about it. I don't know why the idea of home-brew gets me so excited. I suppose it's the same reason I get excited about the notion of milking a cow, or pulling a spud from the ground – it's a little rebellious act of freedom, two fingers to the world of commercialism and mass production. Plus it's 42 cent a pint! Where would you be going?

Home-brew disaster aside, our harvest feast highlighted just how far you can go with this self-sufficiency business – despite the best efforts of slugs and rabbits to eat everything you try to grow, despite the crap Irish weather and our crap, wet soil, despite the complete amateurishness of our efforts, we still managed to turn out a pretty impressive party. On days like that, watching people tucking into your produce, you get all the motivation you need to keep going. It's like a drug. You sit back from the table with a full belly and you could be for-given the arrogance of thinking that it wouldn't take too much additional effort to produce *all* the food that you need, all year round. It's the home-brew that's allowing you this little moment of delusion and, in reality, the only reason that you can pull this off is because it's harvest time and the garden is resplendent with produce and the freezer is full of meat. But still, even in more sober moments I find myself getting very excited about the potential to become more self-sufficient, if not completely so.

There's loads of work to do to get us nearer to that point. In terms of growing vegetables, we have recently put in five new raised beds. I am hoping that these will kill a few birds with one stone – they are high enough that they should, hopefully, deter the rabbits and they will allow us to overcome the fact that our soil is terrible as we can put whatever soil we want in there. They also look nice, which is important when you're growing vegetables in your garden. We made them out of railway sleepers, so the beds are long and narrow, two sleepers long (about 3m), two high (about 1m) and half a sleeper wide. I think they look pretty cool. I know people say that sleepers are a no-no because of all the grease and oil that leaches out of them, but feck the begrudgers, that's all I can say.

It wouldn't take too big a leap for us to become self-sufficient in pork and chicken. We will definitely go again with the pigs. I can't get over how much waste food goes into the bin now that they're gone and I also can't believe how much our meat bill has dropped. We probably need to abandon our faith in Mother Nature when it comes to chicken production – our cock Roger and the broody hens have managed to produce all of two chickens since we got them, which is hardly going to keep us going longterm. In any case, I don't think that the resultant offspring from Roger's criminal trysts with the hens – a Sussex/Rhode Island hybrid – are up to scratch for eating. They just don't grow quick enough and by the time they get to eating weight they are starting to get tough. Perhaps next year I will buy twenty or thirty chicks, the right breed, and build a little pen for them to keep them separate from the hens – and then mercilessly fill the freezer with their delicious, succulent bodies.

Beyond pork and chicken, you're into serious livestock, but I'm not counting myself out on that score just yet. I'm convinced that to keep cows or sheep I would really need to get my hands on some land – not a whole lot of land, but a small field, maybe an acre or two. I'm all for using the space that we have, but our garden is just not big enough for ruminants and it is also too wet. You should see the state of the ground where we kept the pigs – imagine what a cow would do? But if I could get access to a little field, I would seriously love to get a cow and some sheep. I know you're thinking, he's really lost the plot now, but think about it. If you got yourself a nice cow, you would have your own supply of milk and each year she would produce a calf which you could kill off for beef. Simple. You would also have a free, constant supply of the best fertiliser known to man for the vegetable plot. Throw a couple of lambs into the equation and you've got the four major meat types covered. We would probably have to erect some sort of barn for over-wintering, which would be pricey, and there would be a lot of work involved – not least that you would have to spend ten to fifteen minutes twice a day milking the cow, but I'm a routine-freak – I love having a thing that I have to do every day and in any case we are already fairly tied down with the dogs and hens (and the pigs when they were here). It all comes down to the potency of motive – does the desire for self-sufficiency, the desire to know where your food comes from, the desire to get back to the simplicity of the self-supporter lifestyle outweigh the hassle factor? Anyway, all that is for another day. 'Little fields' don't grow on trees, so for now, we have to make do with what we have.

It's very difficult to sum up how I feel about where we are at now, mainly because I don't feel like we have reached the end of a journey. I suppose, with both of us settled into our new jobs for over a year now, I do feel that we've reached a plateau of sorts in our lives, having been on an uphill climb for a long time. It's a more comfortable place to be, certainly, and it allows us a good view of where we've come from, but you get the feeling that there is still some climbing to do up ahead. Professionally, I think we have both reached a level of happiness that I wouldn't have imagined possible three years ago. For my part, I can best explain how I feel about my new job by saying this – last summer I took a week off and on the Monday morning when I was due to go back to work I woke up and found that I was actually looking forward to getting stuck in. That's the difference. That's the feeling that was missing for the past ten years. I lay in bed for a few minutes before I got up and thought about how I would feel right now if I was still in my old job, and I decided I would probably want to roll over and go back to sleep. The other day I drove past the office I had in Waterford and all of a sudden a hundred images of my old life flooded into my mind. All I could think was this: how in the name of God did I stick with it for so long?

Don't get me wrong, I was very lucky to have a good job that paid me well and from which I learned a lot. I don't think I could have got to where I am now without my time in sales and, ironically, a big part of freelancing is being able to sell ideas to editors, so my experience comes in handy. But thankfully, it's a smaller part of my job than it used to be. I feel on top of things in my new job which I never did before. What I work at now is somehow simpler – not easier – but more straightforward. It's hard work coming up with ideas, it's hard work writing them and there's a tremendous responsibility to get the pieces right – but the process itself is straightforward and has a clearly defined beginning, middle and end. There are nights when I am outside in my little office until nine or ten at night, but I never resent that or feel I'm overworked. The weekly deadlines give a certain pressure, but it's an exciting 'let's get the paper out' kind of pressure rather than a debilitating, relentless annual-target kind of pressure.

The pay is absolutely rubbish, of course, comparatively speaking, but I've more or less stopped stressing about all that now. I think I just needed some time to see that we could manage on our new earnings, and that took me about a year or so. Once you reach that point, there is something quite liberating about it. Worry is the most debilitating thing that you can spend your time on – if you allow yourself to slide into worrying about something on any given day, you may as well write off the rest of that day. It's destroyed. And worrying about money is one of the most futile worries of all. If I find myself worrying about money, I ask myself this question: is it costing me anything to live right at this moment? Do I need money to stand or sit here, breathing in and out? The answer is no, I don't. I tell myself: enjoy this moment, the one you are living right now because it's the only moment that you have any control over. Leave the future to look after itself.

I had a very interesting discussion about all this at a wedding recently (you must think I spend my life at weddings). Anyway, the guy I was talking to is an accountant with one of the big-five and he earns mega-money. He was saying that he reckons the amount that you spend will rise to meet whatever you earn. In other words, if you're earning €8,000 a month you will find a way of spending it – you will buy more expensive food, spend more on clothes, have more nights out, more holidays. If you are earning a quarter of that amount, you just won't spend as much on those things. The standard of living in both cases can remain remarkably similar, it's just the cost of 'stuff' in your life increases to match the higher earnings. There's a lot of truth in that. I can't help wondering what the hell *we* did with all the money we earned in our old jobs. I can't say, looking back, that our standard of living was any different – in fact, both of us would agree that it is better now than it was then. And yet we are earning a fraction of what we used to. How does that work?

I think that if you want to make it work on reduced income, there are two things to bear in mind. Firstly you MUST be able to survive day-to-day on your reduced earnings without too much scrimping and saving. If you can't, for whatever reason, then really, you're just going to be miserable. Secondly, you have to downscale your longterm material ambitions. The latter is more difficult than the former. You have to come to terms with the realities that you are never going to live in a big house, drive a nice sports car, buy a big boat, take two holidays and four city breaks a year. In a society where success is defined by exactly these things, that's more difficult than it sounds.

It's really nice to be my own boss, to have the flexibility to decide when and for how long I am going to work. In fact, though I say so myself, I'm the kind of boss I've always dreamed of: understanding, kind, benevolent, tolerant of casual dress and regular tardiness. And the reward, opening up a national newspaper and seeing my name over a feature, still gives me the greatest thrill imaginable. Above all, because of the variety it offers, I can see myself being interested in doing this job in ten years' time, or twenty years' time, or when I'm sixty-five.

So, I guess that means that I can finally say that my new career passes the Wedding Test. I have an answer that I'm proud of for when my dinner companion finally pops the question, as they invariably do:

'So, eh, what do you do?'